CRISPR Chronicles

Navigating the Ethics, Promises, and Perils of Gene Editing

James Ellis

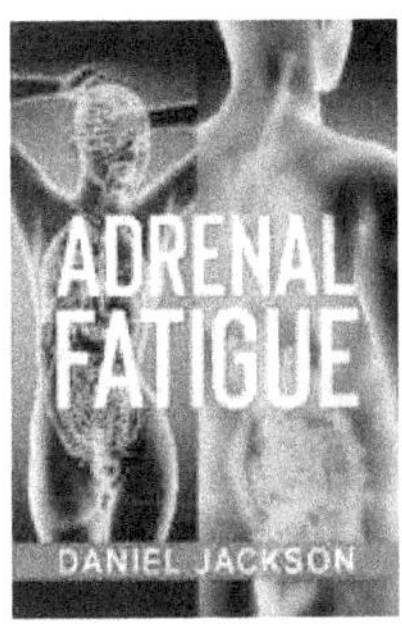

Take a look at more great books available from Rockwood Publishing

... some for FREE!

Just visit the link below:

rockwoodpublishing.co.uk

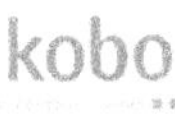

Contents

Introduction: The Dawning of the Genetic Age 1

A brief history of genetic engineering 1

Introduction to CRISPR-Cas9 6

Overview of the Potential of CRISPR Technology 12

Chapter 1: Unraveling CRISPR: Mechanisms and Applications 18

Understanding the Science: How CRISPR Works 18

A Biological Marvel: The Origins of CRISPR 18

The Mighty Scissors: Cas9 19

From Bacterial Defense to Genetic Editing: The Leap 19

The Power of CRISPR-Cas9: Applications and Potential 20

Current Applications: From Laboratories to Clinics 23

Clinical Medicine: A Beacon of Hope 23

Agriculture: A Key to Food Security 24

Conservation and Beyond: The Expanding Horizon 25

The Path Ahead 25

Future Prospects: Where Can CRISPR Lead Us? 28

Healthcare: A Future Free of Genetic Diseases? 28

Food Security and Environment: Toward a Sustainable World 29

Conservation and De-extinction: Rewriting the Story of Life ... 29

Human Enhancement: A New Chapter in Evolution? 30

Chapter 2: The Genetic Revolution: Opportunities and Triumphs 33

Disease Eradication: CRISPR's Potential in Medicine33

Agriculture & Food: Feeding the Planet Sustainably 38

Beyond Health: CRISPR's Role in Biofuels and Materials 43

Chapter 3: Ethics and Accountability: The Line Between Treatment and Enhancement 48

The Morality of Genetic Modification: A Spectrum of Views 48

Editing Embryos: The Question of Germline Modification 53

The Enhancement Debate: From Eliminating Diseases to Designer Babies 58

Chapter 4: Case Studies: CRISPR In Action 64

Saving Lives: Success Stories in Treating Genetic Diseases 64

A Lifesaver for "Bubble Boy Disease" 64

Sickle Cell Disease: From Pain to Potential Cure 65

Huntington's Disease: Slowing the Inevitable 66

Leber Congenital Amaurosis: A Ray of Light 66

Cystic Fibrosis: Breathing Easier 67

Duchenne Muscular Dystrophy: Strength in Science 68

Agricultural Breakthroughs: High-Yield, Disease-Resistant Crops 69

The Dangers: The Case of He Jiankui and Designer Babies
.... 74

Chapter 5: Societal Implications: Access, Equity and
Governance 79

Accessibility and Affordability: Who Gets to Edit?... 79

Governing CRISPR: Regulatory Landscapes Around the
World 82

The Role of Public Perception and Media in Shaping the
CRISPR Debate 84

Chapter 6: A Brave New World: Hypothetical Scenarios
and Their Ethical Implications 88

Scenario Analysis: Utopian and Dystopian Visions .. 88

Ethical Analysis: Applying Ethical Theories to Scenarios
....91

Conclusion: CRISPR and the Future of Humanity ... 94

Recap of the CRISPR Journey 94

Critical Considerations for the Path Forward 96

The Role of Society in Shaping the CRISPR Revolution
.... 102

Introduction: The Dawning of the Genetic Age

A brief history of genetic engineering

Welcome to the dawning of a new age, an era that promises unprecedented advancements and challenges in equal measure - the Genetic Age. This age has been inaugurated by the advent of gene editing, particularly through the revolutionary technique known as CRISPR (Clustered Regularly Interspaced Short Palindromic Repeats), which has transformed our understanding and manipulation of the genetic code.

Before we delve into the intricacies of CRISPR and its vast implications, let's take a step back to understand the historical context and the series of discoveries that have led us to this point. To fully appreciate the strides we've made, it's essential to grasp the brief yet impactful history of genetic engineering.

The Birth of Genetic Understanding

The foundation for genetic engineering was laid over a century ago with the pioneering work of Gregor Mendel, a monk in the 19th century. Known today as the 'Father of Genetics,' Mendel performed a series of experiments with pea plants in his garden, establishing the principles of heredity. He found that certain traits, such as the color or shape of peas, were passed down in predictable patterns, which we now understand as the transmission of genes.

Jumping forward to the 20th century, we come to the discovery of the DNA molecule's structure in 1953. James Watson and Francis Crick, with crucial contributions from Rosalind Franklin and Maurice Wilkins, elucidated the double-helix structure of DNA, the physical and chemical carrier of genetic information. This was a landmark moment in biology, setting the stage for the era of molecular genetics.

First Steps into Genetic Engineering

The first successful instance of what we now call genetic engineering took place in the 1970s. Herbert Boyer and Stanley Cohen combined their expertise in biochemistry and microbiology to create the first recombinant DNA molecule. They achieved this by cutting a specific gene from one organism and inserting it into the DNA of another, paving the way for revolutionary techniques like cloning and the production of genetically modified organisms (GMOs).

In 1978, another significant milestone was reached when scientists used genetic engineering techniques to produce human insulin in bacteria. This marked the first time a genetically engineered product was used to treat a human disease, setting a precedent for the development of countless other therapeutic agents.

The Human Genome and Beyond

As we moved into the 21st century, the spotlight shone on the Human Genome Project, an international effort to map the entire sequence of the human genome. Completed in 2003, the project provided a blueprint for human life at a genetic level, illuminating the function of numerous genes and marking an enormous stride in our understanding of human biology.

The last decade has been particularly transformative, bringing us into the era of CRISPR, a gene-editing tool that has revolutionized the field. Discovered in bacteria as a form of immune defense against viruses, CRISPR systems have been adapted for precise and efficient gene editing in various organisms, including humans.

As we stand at the intersection of biology, technology, and ethics, the Genetic Age asks us to confront profound questions. What does it mean to have the power to alter our own genetic makeup? How will these technologies shape our future? What are the ethical boundaries we must consider?

CRISPR is not just a tool but a symbol of the possibilities and dilemmas that lie ahead. This book aims to guide you through this genetic landscape, examining the ethical, scientific, and societal implications of gene editing. It's an exciting journey, and we invite you to join us as we navigate the complexities of the Genetic Age.

Let's step forward together into this brave new world, for the story of CRISPR is our story – a tale of potential and promise, of challenges and ethical conundrums, of humanity's relentless pursuit of knowledge and mastery over nature.

The CRISPR Revolution

When Jennifer Doudna and Emmanuelle Charpentier first elucidated the mechanism of the CRISPR-Cas9 system in 2012, few could predict the far-reaching implications of their discovery. Derived from a defense mechanism used by bacteria against viruses, CRISPR-Cas9 provided a means to edit genes with unprecedented precision. It's been likened to a pair of molecular scissors, able to cut and paste sections of DNA, enabling scientists to alter genetic sequences at will.

This has dramatically expedited research in various fields, from medicine to agriculture, making gene editing faster, cheaper, and more accurate than ever before. It has fueled hopes for treating genetic diseases, improving crop resilience, and even resurrecting extinct species. The implications are as vast as they are complex, from creating disease-resistant crops that could address food security, to developing gene therapies that could potentially cure genetic disorders like cystic fibrosis or Huntington's disease.

A Doorway to New Possibilities and Ethical Quandaries

Yet, alongside the promise, there are also profound ethical and societal questions to grapple with. What are the long-term effects of gene editing, especially when changes are made to the germ line, affecting future generations? What is the risk of off-target effects, unintended genetic changes that could have unforeseen consequences? Where do we draw the line between curing disease and enhancing human abilities? And who gets to decide?

There are also concerns about the potential for 'designer babies,' where the genetic makeup of an embryo is selected or altered for particular traits, such as intelligence or physical appearance. This opens up a Pandora's box of ethical issues, from exacerbating social inequalities to fundamentally altering our conception of what it means to be human.

Navigating the Crossroads

The progress made in genetic engineering, culminating in the development of CRISPR, represents one of the most transformative scientific advancements of our time. As we traverse the CRISPR crossroads, it's essential that we navigate with caution, ensuring that the benefits of this technology are realized while mitigating potential risks and respecting ethical boundaries.

The Genetic Age isn't a future prospect; it's our current reality. Each new discovery, each scientific paper published, each clinical trial, and indeed each chapter of this book is part of the unfolding narrative of this fascinating epoch. The CRISPR crossroads is a journey into understanding the essence of life, decoding the language of our genes, and envisioning what we, as a species, may become. As we continue to write this story, it is up to us to ensure that the tale is one of enlightenment and progress, rather than recklessness and regret.

Our journey into the Genetic Age is not just about science or technology, but fundamentally about us – our health, our food, our environment, and our society. It's about the choices we make and the world we want to build. And above all, it is about the delicate balance between advancing human knowledge and maintaining our ethical integrity. So, buckle up and prepare for a journey into the heart of the Genetic Age. The road may be complicated, but the exploration is undoubtedly worthwhile.

Introduction to CRISPR-Cas9

On the surface, the acronym CRISPR-Cas9 might sound more like a distant celestial body or a secret code from a spy novel. But for scientists around the world, it's a symbol of revolution. It stands for Clustered Regularly Interspaced Short Palindromic Repeats and the CRISPR-associated protein 9, together forming the most advanced tool in the realm of gene editing. Through this chapter, we will demystify this remarkable technology, explain its functioning, and explore how it has turned the science fiction of gene editing into tangible reality.

CRISPR-Cas9: From Obscurity to Limelight

CRISPR sequences were first discovered in the late 1980s in bacteria, but their purpose was not immediately apparent. It wasn't until the 2000s that researchers figured out they were part of a bacterial defense system against viruses. Bacteria, similar to us, face constant viral threats. To fend off these invasions, they employ a nifty trick: they incorporate snippets of viral DNA into their own genetic code, nestled within the CRISPR sequences.

This library of viral DNA acts as a 'Most Wanted' list. When the same virus tries to invade again, the bacteria produce RNA from the stored viral DNA, which pairs with a protein called Cas9. The RNA guides Cas9 to the matching viral DNA, which the protein then precisely cuts, disabling the virus. In a sense, the bacteria have an adaptive immune system, where they 'remember' past viral foes and know exactly how to fight them off.

Adapting Nature's Tools

This bacterial defense mechanism inspired scientists to envision a new way of editing genes. Could we use the precision of the CRISPR-Cas9 system to target specific genes in other organisms, including humans? The answer, discovered by Emmanuelle Charpentier and Jennifer Doudna in 2012, was a resounding yes. By providing custom-made guide RNAs, scientists can now direct the Cas9 protein to cut any gene at a desired location, marking a step-change in our ability to manipulate the code of life.

The Molecular Scissors

To understand how CRISPR-Cas9 works, imagine the DNA molecule as a vast railway line, extending miles in either direction. This line has numerous 'stations,' representing the genes, each with a specific function. Now, let's say there's a faulty station causing problems in the network — in biological terms, a malfunctioning gene leading to a disease.

To fix this, you need two things: a way to find the faulty station among thousands, and a tool to repair it. The guide RNA is your map; it is designed to match the sequence of the faulty gene. The Cas9 is your toolkit; it cuts the DNA at the identified spot. Once the DNA is cut, the cell's repair machinery kicks in. Scientists can exploit this natural repair process to delete, modify, or replace genes. It's like having a molecular GPS and toolkit, allowing us to navigate and alter the vast genetic landscape with remarkable precision.

The CRISPR Revolution

The implications of CRISPR-Cas9 technology are immense. Its precision, efficiency, and accessibility have democratized gene editing, making it a common tool in laboratories worldwide. It's been used to create disease-resistant crops, study complex genetic disorders, and engineer mosquitoes that can't transmit malaria.

In medicine, the potential of CRISPR-Cas9 is extraordinary. By correcting faulty genes, we could potentially cure genetic disorders like cystic fibrosis, sickle cell disease, or muscular dystrophy. In 2020, two patients with inherited blindness received the first CRISPR-based therapy, marking a significant step towards using gene editing for treating diseases. Clinical trials are also underway to test CRISPR therapies for blood disorders and cancer.

In agriculture, CRISPR can help us create more resilient and productive crops, contributing to food security. It's also being used to study and potentially mitigate the effects of climate change on plant life. Scientists are even exploring the use of CRISPR to drive specific traits through populations, a concept known as a gene drive, which could potentially control or eradicate pest species or disease vectors.

A Balancing Act

While the promise of CRISPR-Cas9 is staggering, it's important to acknowledge the challenges and potential pitfalls. There are concerns about off-target effects, where Cas9 may unintentionally cut other parts of the genome, leading to unpredictable consequences. While advancements are being made to increase the specificity of CRISPR-Cas9, this remains a pertinent issue.

Moreover, the ability to alter the genome so easily brings with it a host of ethical considerations. When, how, and on whom should these technologies be used? Who gets to decide? What about the potential for 'designer babies,' where parents could potentially choose their children's physical and intellectual traits? While the scientific community has been proactive in fostering ethical discussions, these remain complex and contentious issues that society at large must grapple with.

There's also the question of access and equity. Will CRISPR-based therapies be available and affordable to all who need them, or will they exacerbate existing healthcare disparities? As we forge ahead with this technology, it's crucial that we address these issues to ensure that the benefits of CRISPR are shared widely and ethically.

The Road Ahead

The CRISPR-Cas9 technology is a potent symbol of the Genetic Age. It embodies the promise of gene editing, the potential to reshape our world in profound ways. Yet it also encapsulates the complex ethical, social, and scientific challenges that this new age presents.

As we stand on the precipice of this brave new world, we must balance our desire for progress with our responsibility to use these technologies wisely and ethically. We must ensure that our journey into the heart of the genome strengthens, rather than undermines, our shared human values.

The story of CRISPR-Cas9 is more than a tale of scientific discovery. It's a story about us, about our incredible capacity for innovation, and about the choices we make when given unprecedented power over the fabric of life. As we move forward in the Genetic Age, let's ensure that this story is one of caution, courage, and above all, compassion. For in the end, the genome we are editing is not just a sequence of letters, but the blueprint of life itself.

Our journey into the realm of CRISPR-Cas9 is one of awe and anticipation. As we peel back the layers of our genetic blueprint, we are continually reminded of the complexity and beauty of life. While the road may be fraught with ethical dilemmas and scientific challenges, it also presents an opportunity to revolutionize our understanding of life and potentially alleviate human suffering. The Genetic Age, heralded by the power of CRISPR-Cas9, awaits our exploration. So let's embark on this journey with an open mind, a curious spirit, and a cautious optimism for what we might uncover.

Overview of the Potential of CRISPR Technology

The discovery and rapid adoption of CRISPR-Cas9 has brought us into an era of unprecedented potential in the realm of genetic engineering. The advent of this precise, accessible, and relatively affordable tool has transformed the way we approach problems in medicine, agriculture, and beyond, igniting imagination and promise on a global scale. In this chapter, we'll delve into some of the most significant areas where CRISPR technology is anticipated to make a major impact.

Unlocking Therapeutic Potential

In the world of medicine, CRISPR technology is seen as a game-changer. Genetic disorders, which were once considered a life sentence, may now be curable. By precisely editing the DNA sequences responsible for these conditions, we have the potential to eradicate diseases at their source.

One of the most promising areas of CRISPR research is in the treatment of blood disorders like sickle cell disease and beta-thalassemia. Both conditions are caused by single-gene mutations and are, therefore, ideal targets for CRISPR treatment. Early clinical trials have shown promising results, with treated patients experiencing a reduction in symptoms.

Similarly, in the field of oncology, researchers are exploring the use of CRISPR to engineer T cells, a type of immune cell, to better fight against cancers. This method enhances the body's natural defense mechanism and has shown promise in early-stage clinical trials for certain types of cancer.

Yet another exciting frontier is the potential use of CRISPR technology to treat inherited forms of blindness. By editing the faulty gene responsible for a specific type of inherited blindness, scientists hope to restore normal vision. The first clinical trial using this approach has shown encouraging results.

Agriculture and Food Security

Beyond medicine, CRISPR technology holds tremendous promise for global food security. Traditional breeding methods have helped us develop crops with desirable traits, but these techniques are laborious, time-consuming, and lack precision. CRISPR, on the other hand, allows scientists to make specific changes in the plant's genome, leading to improved traits.

Farmers can use CRISPR to develop crops that are more resistant to diseases and pests, reducing the need for chemical pesticides. Similarly, crops can be engineered to withstand harsh environmental conditions such as drought and salinity, a crucial adaptation in the face of climate change.

CRISPR also has the potential to improve the nutritional content of crops. For instance, researchers have used it to increase the levels of essential nutrients in rice, making it healthier for consumption. As we strive to feed a growing global population amidst changing climatic conditions, CRISPR technology could be a crucial tool in ensuring food security.

Biodiversity and Conservation

One of the more unexpected areas where CRISPR shows promise is in biodiversity conservation. By editing the genes of endangered species, we could potentially increase their resilience to disease or changing environments. There is also the tantalizing prospect of 'de-extinction,' where extinct species are brought back to life by editing the genomes of their closest living relatives.

An example is the ongoing effort to resurrect the woolly mammoth. By introducing mammoth genes into Asian elephant cells using CRISPR, researchers hope to create a hybrid animal that can survive in the Arctic tundra, potentially helping to combat climate change.

Industrial and Environmental Applications

CRISPR technology can also be harnessed to develop greener industrial processes. For instance, genetically engineered microbes can be used to produce biofuels, biodegradable plastics, or other useful substances, reducing our reliance on fossil fuels and helping to mitigate climate change.

On the environmental front, CRISPR has been employed to engineer bacteria that can break down plastic waste or to create modified plants that can absorb more carbon dioxide from the atmosphere.

In a similar vein, researchers are exploring the use of CRISPR to address the global crisis of antibiotic resistance. By editing the genes of bacteria, we might be able to develop novel antibiotics or even re-sensitize resistant bacteria to existing ones.

CRISPR and Gene Drives

One of the most radical potential applications of CRISPR technology lies in the creation of 'gene drives.' This technique aims to spread a particular set of genes throughout a population at a rate much faster than traditional inheritance would allow. For example, scientists are investigating the use of gene drives to eradicate or alter populations of disease-spreading mosquitoes. However, the potential ecological impacts of gene drives are largely unknown and thus warrant a great deal of caution.

Navigating the Future of CRISPR Technology

While the potential of CRISPR technology is undeniable, it is also a double-edged sword. The same tools that allow us to cure genetic diseases could also be misused to create 'designer babies.' The prospect of editing the human germline – the DNA that we pass on to our offspring – brings with it significant ethical, moral, and societal challenges.

Furthermore, while CRISPR technology is remarkable in its precision, it is not infallible. Off-target edits, where changes are made to unintended parts of the genome, can occur and could potentially lead to harmful side effects. Therefore, while advancing our capabilities, we must also continuously improve the safety and accuracy of these techniques.

Equitable access to the benefits of CRISPR technology is another key concern. The potential for lifesaving treatments and improved crops should be accessible to all, rather than becoming another source of disparity in a world already divided by wealth and resources.

Despite these challenges, the promise of CRISPR technology is too great to ignore. It is a tool of unprecedented power – one that allows us to shape our world in ways previously only dreamt of in science fiction. With it, we have the potential to cure diseases, feed a growing population, protect our planet, and much more.

As we navigate the future of CRISPR technology, we must do so with a deep sense of responsibility and an unwavering commitment to ethical practice. By embracing both the promises and perils of CRISPR, we can harness its power in a way that benefits all of humanity and the world we inhabit.

Ultimately, the journey of CRISPR technology is only just beginning. As we continue to explore and expand its capabilities, we will likely uncover applications that we cannot even imagine today. The genetic age is upon us, and CRISPR is one of the most potent tools in our arsenal. It provides a window into the blueprint of life, and with it, the potential to rewrite that blueprint for the betterment of all living things. The promise of CRISPR technology is vast; now it is up to us to ensure that promise is fulfilled responsibly and equitably.

Chapter 1: Unraveling CRISPR: Mechanisms and Applications

Understanding the Science: How CRISPR Works

In the realm of science, few discoveries of the 21st century can claim to be as revolutionary as CRISPR-Cas9, a gene-editing technology that is transforming our approach to solving some of humanity's most pressing challenges. But what exactly is this technology, and how does it work? Let's unravel the scientific principles that underpin CRISPR-Cas9.

A Biological Marvel: The Origins of CRISPR

To understand CRISPR, we must first take a journey into the microscopic world of bacteria. Yes, bacteria, those single-celled organisms that live almost everywhere - from the depths of the ocean to the human gut. In the grand chess game of survival, bacteria have a formidable adversary: viruses. To counter viral threats, bacteria have evolved a unique defense mechanism known as CRISPR, which stands for "Clustered Regularly Interspaced Short Palindromic Repeats."

CRISPR is essentially a part of the bacteria's immune system that helps it remember viruses that have previously attacked. When a virus invades, the bacterial cell captures snippets of the viral DNA and stores them in its own genome in a 'memory bank.' If the same virus tries to invade again, the bacteria produce RNA molecules that match the stored viral DNA sequences. These RNAs act as 'most-wanted posters,' helping the cell recognize and destroy the returning enemy.

The Mighty Scissors: Cas9

While CRISPR acts as a memory system, the star player in the mechanism is an enzyme known as Cas9. The Cas stands for "CRISPR-associated," indicating that this enzyme is a part of the CRISPR system. Cas9 acts as the 'scissors' in the CRISPR system – it's the molecule that actually does the cutting of DNA.

Once the RNA molecules recognize the returning virus, they guide the Cas9 enzyme to the viral DNA. The Cas9 enzyme then cuts the DNA at the precise location that matches the RNA sequence. This disables the virus, preventing it from replicating and infecting the bacteria.

From Bacterial Defense to Genetic Editing: The Leap

In 2012, scientists Jennifer Doudna and Emmanuelle Charpentier realized that they could repurpose this bacterial defense mechanism for a completely different use: editing genomes. By designing their own RNA sequences to match a specific location in the DNA, they could direct the Cas9 enzyme to cut any DNA sequence they wanted.

This was a watershed moment. Suddenly, scientists had a precise, easy-to-use, and cost-effective tool to edit genomes. Instead of waiting for evolution to slowly change DNA over millions of years, we could now do it in the lab within days. This technology – known as CRISPR-Cas9 gene editing – has since been adopted worldwide, igniting a revolution in genetic research.

The Power of CRISPR-Cas9: Applications and Potential

The ability to cut DNA at precise locations means scientists can remove, add, or alter specific genes in an organism's genome. The potential applications of this technology are vast and varied.

In medicine, CRISPR-Cas9 is being harnessed to correct genetic disorders, engineer immune cells to fight cancer, and potentially eliminate infectious diseases. In agriculture, it's being used to improve crop resistance to diseases and environmental stressors. And in basic research, CRISPR-Cas9 is enabling scientists to probe the functions of different genes, deepening our understanding of biology.

However, as we explore the power of CRISPR-Cas9, we're also confronting significant ethical and safety questions. What are the risks of making permanent changes to an organism's genome? How do we avoid 'off-target' edits that could have unintended consequences? And perhaps most controversially, should we use this technology to edit the human germline – the DNA that we pass on to our offspring?

Despite these questions and challenges, the scientific community is making significant strides in ensuring that CRISPR-Cas9 technology is used safely, responsibly, and ethically. This journey is not without its hurdles, but the potential benefits are so vast that they warrant our best effort.

In the realm of genetic diseases, CRISPR-Cas9 presents a beacon of hope. With the ability to snip out disease-causing mutations and replace them with healthy DNA sequences, we stand on the brink of a new era in medicine. Hemophilia, cystic fibrosis, muscular dystrophy, and many other conditions caused by single-gene mutations could potentially be eradicated. Even complex diseases like cancer, where multiple genes often play a role, might be more effectively combatted using CRISPR-enhanced immune cells.

In the field of agriculture, CRISPR-Cas9 opens avenues for improving crop resistance against pests and harsh environmental conditions. As climate change makes farming conditions more volatile, these adaptations could be crucial in securing global food supplies.

Beyond these immediate applications, the potential for CRISPR-Cas9 technology continues to expand as scientists innovate and explore new uses. Conservation biologists are investigating its potential for protecting endangered species and restoring biodiversity. Others are exploring its use in tackling environmental challenges, such as developing biofuels or engineering bacteria that can break down plastic waste. And some scientists are even discussing the possibility of using CRISPR-Cas9 to bring extinct species back to life.

As we continue to unravel the full potential of CRISPR-Cas9, we must always be guided by a commitment to safety, responsibility, and ethical practice. As with any powerful technology, the path ahead is both exciting and fraught with challenges. But with care and vigilance, we can harness the power of CRISPR-Cas9 to address some of the most pressing challenges facing our world today.

So, we stand at the dawn of a new era in genetic engineering, guided by the beacon of CRISPR-Cas9. It's a technology rooted in the fundamental principles of biology, yet one that offers us the power to shape the very blueprint of life. As we navigate the road ahead, let's ensure that we harness this power with wisdom, responsibility, and a deep respect for the marvels of life's complexity. The story of CRISPR-Cas9 is still being written, and we all have a role to play in shaping its course.

Current Applications: From Laboratories to Clinics

As we delve further into the applications of CRISPR-Cas9, it's essential to understand that its reach extends far beyond laboratory walls. This game-changing technology is already making waves in clinics, fields, and conservation areas worldwide, presenting novel solutions to some of humanity's most pressing challenges.

Clinical Medicine: A Beacon of Hope

In the realm of clinical medicine, CRISPR-Cas9 is often hailed as a potential game-changer, and rightly so. The technology's power to precisely edit genomes opens up promising avenues for tackling a wide range of diseases.

Take, for instance, genetic disorders. These conditions, caused by mutations in specific genes, have long been considered largely untreatable. However, with CRISPR-Cas9, we now have the potential to correct these mutations directly. In 2020, the first CRISPR-based therapy entered clinical trials for patients with sickle cell disease and beta-thalassemia. By editing the patient's stem cells to produce healthy red blood cells, doctors hope to offer a cure rather than just managing symptoms.

Cancer, too, is in CRISPR's crosshairs. Innovative treatments are being explored, such as engineering T-cells - a type of immune cell - to better recognize and destroy cancer cells. In 2020, a team of scientists at the University of Pennsylvania reported the first successful use of CRISPR-Cas9 in three cancer patients, marking a crucial milestone in the fight against this dreaded disease.

Agriculture: A Key to Food Security

Beyond the realm of human health, CRISPR-Cas9 is also revolutionizing the field of agriculture. By allowing precise edits to crop genomes, CRISPR can help us develop plants that are more resistant to pests and diseases, withstand harsh environmental conditions, or even possess improved nutritional profiles.

For instance, scientists have already used CRISPR to develop a variety of wheat that is resistant to powdery mildew, a major cause of crop loss worldwide. In another example, researchers have used the technology to modify rice plants to better withstand flooding - a trait that could prove vital as climate change intensifies.

CRISPR-Cas9 can also address nutritional deficiencies. A team of scientists at the Swiss Federal Institute of Technology used CRISPR to increase the iron and zinc content in a variety of rice. Such biofortified crops could play a crucial role in combating malnutrition across the globe.

Conservation and Beyond: The Expanding Horizon

The potential applications of CRISPR-Cas9 extend even beyond human health and agriculture. Conservation biologists are exploring the use of CRISPR to protect endangered species, restore ecosystems, and even potentially resurrect extinct species. In one ambitious project, scientists are attempting to use CRISPR to engineer elephant cells with woolly mammoth genes, with the aim of reintroducing these 'mammoth-elephants' to the Arctic to slow down the melting of permafrost.

Furthermore, some scientists are exploring CRISPR's potential in combating environmental pollution. Researchers are engineering bacteria with CRISPR to break down plastic waste, potentially offering a biological solution to this pressing environmental issue.

The Path Ahead

The examples above provide a snapshot of how CRISPR-Cas9 is already being applied to real-world problems. However, it's crucial to remember that we're still in the early days of this technology. As we continue to improve our understanding and control of CRISPR-Cas9, its range of applications will likely expand even further.

As we navigate this brave new world of gene editing, we must tread carefully. The power of CRISPR-Cas9 comes with ethical and safety questions that we must diligently address. Off-target effects, where unintended regions of the DNA are altered, remain a critical concern that researchers are tirelessly working to minimize. In clinical settings, the potential long-term impacts of gene editing on patients also require careful, ongoing study.

Ethical questions are equally, if not more, challenging. For instance, should we use CRISPR to edit the human germline, the DNA that we pass on to our children? Such edits would not only affect the individual but also their descendants and potentially the entire human gene pool.

The implications are enormous, and the debate is ongoing. Similar ethical questions arise in other applications. In agriculture, how do we balance the potential benefits of genetically edited crops against concerns about biodiversity and the unknown long-term effects of these alterations? In conservation, is it right to manipulate the genes of endangered species or even bring extinct species back to life? These questions don't have easy answers, but they are crucial to consider as we forge ahead with this transformative technology.

As we unravel the potential of CRISPR-Cas9, it is also essential to emphasize the importance of transparency and public engagement. The decisions we make about how to use, regulate, and control this technology will shape our future. Therefore, these decisions must be made collectively, involving not just scientists and policymakers, but society as a whole.

Despite these challenges, the promise of CRISPR-Cas9 remains undimmed. As we stand on the cusp of a new era in genetic engineering, it's clear that this technology holds immense potential to transform our world. Whether it's curing previously untreatable diseases, ensuring food security, protecting our biodiversity, or tackling pollution, CRISPR-Cas9 could play a vital role in addressing some of humanity's most pressing challenges.

However, the road to realizing this potential is not a straightforward one. It's a path that requires us to navigate complex scientific, ethical, and societal terrain. But with care, responsibility, and a commitment to open dialogue, we can harness the power of CRISPR-Cas9 to build a better future.

As we continue to explore the applications of CRISPR-Cas9, from laboratories to clinics, fields to conservation areas, let's remember that this journey is not just about harnessing a powerful technology. It's also about defining who we are as a society, what we value, and how we envision our future. The CRISPR revolution is more than a scientific story; it's a human story, a story about our collective aspiration to improve life on this planet. And that's a narrative we all have a role in shaping.

Future Prospects: Where Can CRISPR Lead Us?

As we stand at the crossroads of a genetic revolution, with CRISPR-Cas9 as our guide, the horizon is radiant with promise. This technology, as transformative as it already is, is just getting started. As we refine our understanding and improve our control over CRISPR, its potential applications continue to broaden and deepen. The future of CRISPR, while still largely unwritten, hints at possibilities that just a few years ago, were the stuff of science fiction.

Healthcare: A Future Free of Genetic Diseases?

One of the most significant areas of potential for CRISPR lies in healthcare. We've already touched on its current applications in treating genetic disorders and cancer, but looking forward, the technology could transform medicine even more fundamentally.

Imagine a world where a simple blood test and a course of personalized gene therapy could cure ailments that have plagued humanity for millennia. From Alzheimer's and Parkinson's to diabetes and heart disease, numerous conditions have a genetic component that CRISPR could potentially address. In the realm of infectious diseases, CRISPR could also be used to engineer immunity to pathogens like HIV, or to rapidly develop targeted therapies for emerging diseases.

Food Security and Environment: Toward a Sustainable World

As the global population continues to rise and climate change threatens our food security, CRISPR could be a critical tool in ensuring a sustainable future. By enabling the creation of crops that are more resilient, nutritious, and productive, CRISPR could help us feed a growing world without further straining our planet's resources.

Moreover, CRISPR also holds potential for addressing environmental challenges. We've already seen early attempts to engineer bacteria to break down plastic waste. Looking forward, it's possible we could use CRISPR to tackle other forms of pollution or even directly combat climate change, for example, by engineering plants or algae to sequester more carbon dioxide.

Conservation and De-extinction: Rewriting the Story of Life

In the realm of conservation, the potential of CRISPR is vast. We've already mentioned the idea of using CRISPR to protect endangered species or restore ecosystems. But what about reversing extinction?

The notion of de-extinction – bringing extinct species back to life – was once pure fantasy. But with CRISPR, it's now within the realm of possibility. For instance, scientists are exploring the possibility of using CRISPR to reintroduce the passenger pigeon, once a common sight in North America before it went extinct in the early 20th century.

However, de-extinction comes with its own set of ethical and ecological considerations. Would these revived species be able to survive and reproduce in today's world? How would they impact current ecosystems? These are some of the questions we'll need to grapple with as we explore this frontier.

Human Enhancement: A New Chapter in Evolution?

Perhaps the most controversial potential application of CRISPR is in human enhancement. Here, we're not just talking about curing diseases, but enhancing our natural abilities.

Could we use CRISPR to make ourselves smarter, stronger, or more resistant to aging? Could we engineer our children to be taller, more athletic, or musically gifted? Could we even introduce entirely new capabilities, like the ability to see in ultraviolet light or resist radiation, as some organisms can?

These scenarios raise profound ethical and societal questions. If such enhancements become possible, who would have access to them? Could we end up in a world where the genetically enhanced form an elite class, widening the gap between the 'haves' and the 'have nots'?

These are complex issues that we'll need to confront as a society.

As we peer into the future of CRISPR, it's clear that the potential is enormous, but so are the challenges. It's crucial that we navigate this path with care, considering not just the scientific and technical aspects, but also the ethical, societal, and ecological implications.

The future of CRISPR is not a fixed destination, but a journey. It's a journey that requires us to ask hard questions, to strive for transparency, and to engage in open, inclusive dialogue.

Despite these challenges, the promise of CRISPR remains vast. From revolutionizing medicine and agriculture to rewriting the story of life itself, CRISPR could reshape our world in profound ways. But it's essential to remember that the power of CRISPR is not just in the technology itself, but in how we choose to use it.

In the end, the future of CRISPR is not just about harnessing a powerful technology, but about making wise choices that reflect our values and aspirations. It's about exploring the possibilities while respecting the limits, about pushing the boundaries of what's possible while considering the consequences.

As we stand at the crossroads of this genetic revolution, the path ahead is radiant with promise, but it is also fraught with challenges. As we step into this brave new world of gene editing, let's do so with our eyes wide open, our minds engaged, and our hearts committed to building a better, more equitable, and more sustainable world.

The story of CRISPR is not just a scientific saga; it's a human story, a story about our collective potential to shape the future. And that's a narrative we all have a role in writing. So as we look ahead to the future of CRISPR, let's remember that this is not just a journey of science, but a journey of humanity, a journey that we're all a part of.

Together, let's make it a story of hope, wisdom, and progress.

Chapter 2: The Genetic Revolution: Opportunities and Triumphs

Disease Eradication: CRISPR's Potential in Medicine

From its humble origins in bacterial immune systems, CRISPR has quickly become one of the most potent tools in our medical arsenal, promising to usher in a new era in disease treatment and prevention. This technology has dramatically expanded our ability to alter the DNA within living cells, opening the door to unprecedented interventions at the most fundamental level of biology.

Here, we'll explore some of the ways in which CRISPR is poised to transform medicine, potentially revolutionizing our approach to a wide array of illnesses from genetic disorders to cancer.

Targeting Genetic Diseases at their Source

Perhaps the most straightforward application of CRISPR in medicine lies in the treatment of genetic disorders. These are diseases that result from specific mutations or errors in our DNA, which cause cells to function abnormally. The list of such diseases is long, from cystic fibrosis and Huntington's disease to certain forms of blindness and deafness.

Traditionally, treating genetic disorders has been challenging. Most treatments aim to manage the symptoms of the disease rather than addressing the underlying genetic cause. CRISPR, however, gives us the potential to correct these genetic errors directly. By introducing a CRISPR system into the patient's cells, we can effectively 'edit' their DNA, replacing the disease-causing mutation with a healthy version of the gene.

The potential impact of this approach is enormous. Imagine a world where children born with genetic disorders could be treated with a single injection, replacing a lifetime of chronic disease with a healthy, normal life. While such a reality is still a few years away, early clinical trials using CRISPR to treat genetic diseases have shown promising results.

For example, in a trial to treat a form of inherited blindness, researchers used CRISPR to correct a mutation in the retinal cells of patients, successfully restoring some level of vision. In another trial, researchers used CRISPR to treat beta thalassemia and sickle cell disease, genetic disorders affecting the blood. Patients who received the treatment saw significant improvements in their symptoms, with some individuals even appearing to be cured.

Revolutionizing Cancer Treatment

In addition to treating genetic disorders, CRISPR also holds immense potential in the fight against cancer. Many forms of cancer are driven by specific genetic mutations within cells. With CRISPR, we have the ability to target these cancer-causing mutations directly, either by correcting the mutation or by disabling the cancer cell's ability to proliferate.

Moreover, CRISPR could dramatically enhance our ability to harness the body's immune system to fight cancer. A promising new approach in cancer treatment involves modifying a patient's immune cells, specifically a type called T-cells, to recognize and kill cancer cells. This method, known as CAR-T cell therapy, has proven highly effective in some types of cancer, but it has limitations.

CRISPR could overcome some of these limitations. By using CRISPR to edit the genes of T-cells, we can potentially create a more robust, flexible, and precise cancer-fighting force within the body. Early trials using CRISPR-modified T-cells have shown promise, indicating that this could be a powerful new tool in our cancer-fighting arsenal.

Preventing Disease Before it Starts

In addition to treating existing diseases, CRISPR also has the potential to prevent disease before it even begins. One avenue for this is through what's known as germline editing, where changes are made to the DNA of sperm, eggs, or early embryos. These changes would then be inherited by future generations, potentially eradicating certain genetic diseases from a family line altogether.

Moreover, CRISPR could also be used to confer resistance to infectious diseases. For example, researchers are exploring how to use CRISPR to engineer immune cells with resistance to HIV, potentially providing a cure for this devastating disease. In addition, the same principle could be applied to other infectious diseases, offering a powerful new tool in our ongoing battle against pandemics.

It's important to note, however, that the use of CRISPR for disease prevention, especially through germline editing, raises significant ethical and societal questions. Who gets to decide what constitutes a disease worth preventing? Could the technology be misused to select for non-medical traits like height or intelligence? And who would have access to such technologies? These are complex questions that we'll need to confront as we navigate the future of CRISPR in medicine.

A Word of Caution

While the potential of CRISPR in medicine is enormous, it's important to remember that this is still a very new technology. Many of the applications we've discussed are still in the early stages of development and testing. There are many technical challenges to overcome, and the safety and efficacy of these treatments need to be thoroughly assessed.

Moreover, we must also grapple with the ethical, societal, and regulatory implications of this technology. How do we ensure that CRISPR-based treatments are safe and effective? How do we balance the potential benefits of gene editing with the risks and ethical considerations? And how do we ensure that the benefits of this revolutionary technology are accessible to all, not just a privileged few?

Navigating these challenges will require not just scientific and medical expertise, but also robust public dialogue, thoughtful policy-making, and careful ethical consideration.

As we stand on the brink of a new era in medicine, it's clear that CRISPR holds immense potential. From treating genetic disorders and cancer to preventing disease before it starts, this technology could transform our approach to health and disease, bringing us one step closer to a world free of some of the most devastating illnesses.

But with great power comes great responsibility. As we chart the future of CRISPR in medicine, we must navigate carefully, balancing the promise of new treatments with the ethical, societal, and technical challenges this technology presents.

The genetic revolution is here, and with it, a new era in medicine. As we embark on this journey, let's do so with our eyes wide open, embracing the promise of CRISPR while grappling honestly and earnestly with its challenges.

Agriculture & Food: Feeding the Planet Sustainably

In an era marked by climate change, a burgeoning global population, and an increasing demand for sustainable and nutritious food, the agricultural sector faces significant challenges. However, the same technological innovation that's poised to transform medicine — CRISPR — also offers promising solutions for these pressing issues. Through its precision and versatility, CRISPR could help create a more sustainable and resilient food system. Let's delve into how this genetic technology can assist us in feeding the planet sustainably.

Enhancing Crop Productivity and Resilience

The backbone of global food security is agricultural productivity: our ability to produce enough crops to feed the world's population. However, agricultural productivity is threatened by various factors including diseases, pests, and increasingly, climate change. Rising temperatures, shifting precipitation patterns, and extreme weather events can devastate crop yields.

This is where CRISPR comes in. By precisely editing the genes of crops, we can potentially make them more resilient to these threats. For example, scientists are using CRISPR to develop strains of wheat that are resistant to powdery mildew, a destructive fungal disease. Similarly, researchers are working on creating varieties of rice and corn that can withstand drought or flooding.

This technology isn't just about survival; it's also about performance. CRISPR can potentially enhance crop productivity by tweaking genes that control plant growth and development. With these modifications, crops can be engineered to have higher yields, grow in denser configurations, or mature more quickly, all of which can increase overall agricultural output.

Reducing the Environmental Impact of Agriculture

CRISPR doesn't just offer the potential to produce more food; it could also help us produce food more sustainably. Modern agriculture can take a heavy toll on the environment, from deforestation and soil degradation to water pollution and greenhouse gas emissions. CRISPR presents a powerful tool to alleviate these environmental pressures.

For instance, consider nitrogen. Many crops require large amounts of nitrogen fertilizer for optimal growth, but producing and applying these fertilizers contributes significantly to greenhouse gas emissions and water pollution. Some researchers are using CRISPR to engineer crops that can more efficiently absorb and utilize nitrogen, reducing the need for artificial fertilizers.

Pest management is another area where CRISPR can make a difference. Traditional pest control methods often rely heavily on chemical pesticides, which can have harmful environmental and health impacts. But with CRISPR, we can potentially engineer crops that are resistant to pests, reducing the need for these harmful chemicals.

Improving Nutritional Content

Beyond quantity and sustainability, there's also the issue of quality. Despite producing more food than ever before, many parts of the world still grapple with nutritional deficiencies. CRISPR offers a way to address this issue by enhancing the nutritional content of crops.

For example, scientists are using CRISPR to increase the levels of essential nutrients in staple crops. From iron-rich rice to vitamin A-enhanced bananas, these 'biofortified' crops could go a long way towards addressing malnutrition. By making our crops healthier, we can make our global food system not just more sustainable, but also more nourishing.

Navigating the Challenges

As promising as these applications are, it's important to acknowledge the hurdles and complexities. For one, while CRISPR is more precise than previous genetic modification methods, it's not perfect. Unintended modifications could potentially occur, and their impacts on ecosystems and human health need to be fully understood.

Public acceptance and regulatory oversight are other significant challenges. People are understandably wary about genetically modified organisms (GMOs), and the integration of CRISPR-modified crops into our food system will require transparent communication and rigorous safety testing.

As we face the immense challenge of feeding a growing population in a changing climate, CRISPR offers powerful new tools to enhance the sustainability, resilience, and nutritional value of our crops. From developing disease-resistant strains to reducing agriculture's environmental footprint and improving the nutritional profile of crops, this technology holds significant promise for the future of food.

But with this potential comes a responsibility to navigate carefully. The genetic revolution is reshaping agriculture, and we must ensure that we understand the implications, from the ecological impacts of genetically modified crops to the societal implications of their deployment. We need transparent, evidence-based public discussions and robust regulatory oversight to guide the integration of CRISPR technology into our food system.

Moreover, CRISPR is not a silver bullet. It's a powerful tool, yes, but addressing global food security will require a multi-faceted approach. It will take concerted efforts across multiple sectors, from agriculture and environment to health and policy, to build a truly sustainable and equitable food system.

As we navigate this genetic revolution in agriculture, let's do so with an eye towards the future. With CRISPR, we have the chance to shape the future of food – a future where everyone has access to nutritious, sustainable, and resilient food systems. However, to realize this vision, we need to balance the promise of genetic technology with thoughtful consideration of its risks and implications.

In the end, the goal isn't just to produce more food, but to produce better food, and to do it in a way that respects our planet and its future generations. With CRISPR, we have an exciting new tool to help us reach this goal. But it's up to us to use it wisely, and that's a responsibility we should all take to heart.

Beyond Health: CRISPR's Role in Biofuels and Materials

While the applications of CRISPR-Cas9 in health and agriculture are frequently discussed, the potential of this groundbreaking technology extends much further. Remarkably, CRISPR holds the power to reshape the way we produce energy and materials, contributing to a sustainable and efficient future. Let's delve into these intriguing prospects.

A Greener Fuel Future with CRISPR

The search for renewable and cleaner energy sources is one of the most pressing challenges of our time. Biofuels - fuels derived from living matter - offer an exciting solution. However, the production of biofuels is often associated with sustainability concerns, including land use issues and the overall efficiency of biofuel-producing organisms. This is where CRISPR comes in.

By using CRISPR to edit the genes of microorganisms such as bacteria and algae, we can potentially improve their ability to convert sunlight, CO_2, and other resources into biofuels. For instance, scientists have used CRISPR to tweak the metabolic pathways of E. coli bacteria, boosting their production of biofuels such as ethanol and biodiesel.

Similarly, researchers are using CRISPR to improve algae's ability to produce lipids - oily substances that can be converted into biofuel. Algae have several advantages over traditional biofuel crops: they grow faster, can thrive in a variety of environments, and don't compete with food crops for land.

In addition, CRISPR technology could also be used to enhance the conversion of biomass, such as agricultural waste, into biofuels. This represents a way to turn what is often considered waste into a valuable resource, further improving the sustainability of our energy production.

Revolutionizing Material Production

Beyond biofuels, CRISPR also holds significant promise for the production of sustainable materials. Plastics, for instance, are a ubiquitous part of modern life, yet their environmental impact is significant. They're made from non-renewable petroleum resources, and they persist in the environment for hundreds of years.

Bio-based, biodegradable plastics could provide a more sustainable alternative. CRISPR can potentially help here by optimizing the ability of microorganisms to produce biodegradable plastics, such as polyhydroxyalkanoates (PHAs). By modifying the relevant metabolic pathways in these microorganisms, we could enhance the efficiency and economic viability of biodegradable plastic production.

Furthermore, the potential of CRISPR in material science extends beyond plastics. The technology could also be used to produce stronger, more durable materials by modifying the genes of organisms that produce substances like spider silk, which is stronger than steel on a per-weight basis.

Navigating the Challenges

Despite the exciting potential, applying CRISPR technology in these sectors comes with its own set of challenges. Achieving significant yield improvements for biofuels or material production may require complex changes to an organism's metabolic pathways. This necessitates a deep understanding of these pathways and the interactions between them, something scientists are still working to fully grasp.

There are also ecological and biosecurity considerations. The release of genetically modified organisms into the environment, even if it's for beneficial purposes, needs to be carefully evaluated to prevent unintended ecological impacts.

Regulatory and societal acceptance is another hurdle. As with any application of genetic engineering, transparent communication about the benefits, risks, and safeguards is crucial to gain public trust and acceptance.

The applications of CRISPR extend well beyond the realms of health and agriculture. Its potential to revolutionize the production of biofuels and materials opens up exciting possibilities for a more sustainable future. As we strive to reduce our carbon footprint and move towards more renewable resources, CRISPR could prove an invaluable ally.

Navigating the path forward will require careful balancing of benefits, risks, and ethical considerations. The guiding principle should be responsible innovation. As with all technology, the goal of deploying CRISPR in the realms of biofuels and materials should not only be about pushing the boundaries of what is technically possible, but also about ensuring that the technology is used in a manner that is safe, ethical, and beneficial for society at large.

The prospect of bioengineered solutions to our energy and material needs is both thrilling and daunting. However, our capacity to transform the living world around us, even at the molecular level, is a testament to human ingenuity and the remarkable tools that science has put in our hands. Yet, the use of these tools should always come with a humility and respect for the complexities of nature. The story of CRISPR is ultimately not just about rewriting the code of life but also about learning to navigate the ethical, ecological, and societal implications of this power.

This journey takes us far beyond the realm of science, into the realm of values, vision, and collective decision-making. It forces us to ask, not just what we can do with the power of genetic engineering, but also what we should do.

As we look towards the future, let's ensure that our exploration of CRISPR's potential in biofuels and materials is guided by a vision of sustainability and the welfare of our planet. This path may not be without its challenges, but with careful consideration, the potential rewards - a cleaner environment, sustainable energy, and a more resilient economy - are immense.

In the end, CRISPR isn't just a scientific tool. It's a symbol of our remarkable capacity to understand and shape the living world. How we use this capacity will help define our future. We have an opportunity - and a responsibility - to use it wisely. The future of biofuels, materials, and much more, may depend on it.

Chapter 3: Ethics and Accountability: The Line Between Treatment and Enhancement

The Morality of Genetic Modification: A Spectrum of Views

With great power comes great responsibility. The phrase might be a comic book cliché, but it's a sentiment that rings strikingly true in the world of genetic editing. As we've seen in the previous chapters, the CRISPR-Cas9 system, with its ability to make precise changes to the genetic code, offers a host of incredible opportunities in medicine, agriculture, and beyond. But alongside these promises are profound ethical questions that society must grapple with.

The ethics of genetic modification aren't a monolith; opinions exist along a spectrum that spans from strong opposition to enthusiastic endorsement, each presenting compelling arguments.

The Concerned Traditionalists

At one end of this spectrum are those who express deep ethical reservations about genetic modification. They view the genetic code as a sacred part of the natural order, something we shouldn't meddle with. Some of their concerns stem from a belief in the intrinsic value of the natural world and a respect for its complexities, which human intervention might disrupt.

Others question whether it's fair or ethical to make irreversible changes to a person's genome — or the genome of an entire species — without their consent. And some worry about a potential slippery slope, where what begins as medical treatment evolves into genetic enhancements, potentially leading to societal divisions or even a new form of eugenics.

The Cautious Optimists

Further along the spectrum are the cautious optimists. This group acknowledges the transformative potential of CRISPR technology but also recognizes the need for robust ethical guidelines and regulation.

They often advocate for a careful, case-by-case approach to genetic modification. For them, the moral acceptability of CRISPR usage depends on the specific context: the nature of the condition being treated, the availability of alternative therapies, the potential risks and benefits, and the extent to which the person affected (or their guardians) have given informed consent.

Cautious optimists also stress the importance of transparency, public engagement, and oversight. They argue that decisions about genetic modification shouldn't be made solely by scientists or corporations, but should involve a broad societal dialogue that takes into account diverse perspectives.

The Progressive Futurists

At the other end of the spectrum, we find the progressive futurists. This group embraces the potential of genetic modification and views it as a powerful tool for human betterment.

The futurists may argue that, rather than violating nature, genetic modification is simply the next step in human evolution – a step we're privileged to consciously take. They emphasize the potential benefits of CRISPR technology, from curing devastating genetic diseases to enhancing human capacities and resilience.

Some even propose a moral imperative to use genetic editing. If we have the power to eliminate diseases, enhance longevity, or reduce suffering, don't we have a responsibility to do so?

Navigating the Spectrum

Where should society land on this spectrum? It's a question with no easy answer. The morality of genetic modification is a deeply complex issue, intertwining scientific facts, philosophical beliefs, personal values, and societal norms. It's a conversation that needs to be inclusive, nuanced, and ongoing, evolving alongside our understanding and application of the technology.

It's important to note that this spectrum isn't about 'right' or 'wrong' viewpoints. Rather, it reflects the richness of perspectives in our global society. As we navigate the ethical landscape of genetic modification, we need to respect this diversity of views and strive for dialogue and understanding.

The Delicate Line Between Therapy and Enhancement

One of the most contentious issues in the ethics of genetic modification is the distinction between therapy and enhancement. Therapeutic uses aim to treat or prevent diseases, while enhancements seek to improve human abilities beyond their typical range.

On the surface, the distinction seems straightforward. Treating cystic fibrosis or muscular dystrophy with CRISPR could be seen as a therapeutic use, while using it to increase intelligence or athletic ability might be seen as enhancement. But the line between the two is not always clear-cut. For instance, would using CRISPR to boost the immune system to superhuman levels be classified as therapy or enhancement?

Furthermore, this distinction raises a challenging question: If it's ethically acceptable to use CRISPR to correct genetic 'faults,' might it also be acceptable — or even desirable — to use it to improve upon 'normal' human traits?

Unintended Consequences: The Butterfly Effect

The complexity of biological systems implies that genetic modifications, even when well-intentioned and seemingly minor, can have unexpected and far-reaching effects. A change in one part of the genome might have knock-on effects elsewhere, leading to unforeseen health risks.

Moreover, if we modify the genes of plants, animals, or entire ecosystems — as some propose for controlling disease-carrying mosquitoes or rescuing endangered species — we could unwittingly upset ecological balances with potentially drastic consequences.

Justice and Access: The Ethics of Distribution

Another ethical issue to be tackled is that of justice and access. Like many cutting-edge medical technologies, CRISPR has the potential to be expensive. If only the wealthy can afford gene therapies or enhancements, we could see a deepening of existing social inequalities.

Furthermore, there's a risk of creating a genetic divide, where people with access to genetic modification could enhance their health, longevity, and abilities, leaving those without access even further behind.

Towards an Ethical Framework for Genetic Modification

As we grapple with these and other ethical challenges, it's clear that the scientific community and society at large have a crucial role to play in shaping the path forward. This involves ongoing dialogue, thoughtful regulation, and constant vigilance to ensure that the immense power of genetic modification is used wisely and ethically.

The conversation about the ethics of genetic modification is a journey — and like all journeys, it's as much about the path we take as the destination we reach. As we navigate the CRISPR crossroads, we'll continue to explore these questions, and look ahead to the emerging ethical, legal, and social implications of this revolutionary technology.

What's certain is that the genetic revolution is not just a scientific endeavor; it's a deeply human one, fraught with promises and perils, challenges and opportunities. How we wield this power will say much about who we are — and the kind of world we want to build.

Editing Embryos: The Question of Germline Modification

As we peer over the precipice of the genetics era, one question looms large and controversial above the rest: Should we use CRISPR to modify human embryos? This question gets to the heart of what it means to tinker with life at its most elemental level, and the implications stretch far beyond the individuals immediately affected.

What is Germline Modification?

To understand the stakes, we need to first define what we mean by germline modification. The human body consists of two basic types of cells: somatic cells, which make up most of the body, including skin, organs, and blood; and germline cells, which are the reproductive cells (eggs and sperm) that pass on genes to the next generation.

When we talk about germline modification, we're referring to changes made to the DNA of eggs, sperm, or embryos. These changes can affect the entire organism that grows from the edited cell, and more importantly, these modifications are inheritable. That means the changes can be passed down to subsequent generations.

The Promise of Germline Modification

The potential benefits of germline modification are immense. It could allow us to eliminate genetic diseases from a family line forever, meaning that future generations would no longer be at risk. Imagine wiping out the hereditary basis of cystic fibrosis, Huntington's disease, or certain types of hereditary cancer. The power to erase such conditions from our genetic heritage is the kind of medical breakthrough humanity has dreamt of for centuries.

Germline editing could also have broader applications in the realm of public health. For instance, by editing the genes associated with susceptibility to diseases like HIV, we could create generations of humans naturally resistant to such viruses, revolutionizing our approach to combating such diseases.

The Perils of Germline Modification

However, this power does not come without significant risks and ethical dilemmas. Firstly, our knowledge of the human genome and the long-term effects of genetic modification is still incomplete. Every gene interacts in complex ways with other genes and the environment. Making changes in one area could have unforeseen effects in others.

Moreover, germline modifications, once made, are permanent and inheritable. Mistakes cannot simply be reversed. They are propagated through the generations, potentially affecting scores of descendants. The responsibility of wielding such power is, therefore, enormous.

The Ethical Minefield

From an ethical standpoint, germline modification raises challenging questions. It's one thing to consent to gene therapy as an individual, but can we consent on behalf of our future children, grandchildren, and generations yet unborn? Some argue that it's not our place to make such decisions for those who can't yet voice their consent.

And then there's the concern about the potential for misuse. If we accept that it's permissible to edit embryos to eliminate disease, could we soon see parents opting for enhancements, such as greater intelligence or physical prowess, for their future children? Would we then be opening the door to a world of 'designer babies', with all the ethical and societal implications that would entail?

The Global Debate

The international community is sharply divided on this issue. While some countries have laws explicitly prohibiting germline modification in humans, others have yet to regulate the practice. The scientific community is also wrestling with these issues. In 2015, a group of Chinese researchers sparked global controversy when they announced the first use of CRISPR to edit non-viable human embryos.

Following the uproar, the National Academies of Sciences, Engineering, and Medicine in the United States convened an international summit to discuss the ethical issues surrounding human genome editing. The consensus was that while it would be "irresponsible" to proceed with any clinical use of germline editing until safety and efficacy issues are resolved, the research involving human embryos is permissible, but should be tightly regulated.

This delicate balance was tipped again in 2018 when a Chinese scientist, He Jiankui, announced the birth of the world's first gene-edited babies, twin girls whose genes he claimed to have altered to make them HIV-resistant. The global scientific community was largely horrified, and the incident resulted in calls for a global moratorium on editing human embryos intended for pregnancy.

The Path Forward

The case of He Jiankui demonstrates that the technology for germline editing is here, and the stakes are high. As we stand on the cusp of this new genetic age, it's clear that global consensus and regulations are needed. Such consensus, however, is hard to achieve. Our cultural, religious, and personal beliefs inform our attitudes towards such profound interventions in human life, and these beliefs vary widely across the globe.

The conversations are tough but necessary. They should be inclusive, involving not only scientists, ethicists, and policymakers, but also the public. After all, the outcomes of these debates will impact all of society. Education and engagement are key to ensure that the public understands the science, the potential benefits, and the risks.

At the same time, the scientific community needs to continue its research to enhance our understanding of our complex genetic machinery. Caution should be the watchword here. We need to recognize that with the power to alter the human genome comes great responsibility.

A Crossroads of Humanity

Germline modification represents a crossroads for humanity. The decisions we make now will resonate far into the future, shaping not just the destiny of our species, but also our values, our societies, and our conception of what it means to be human.

The questions we face are challenging and fraught with uncertainty. Yet, in wrestling with these questions, we have the opportunity to reflect on our shared values, consider the kind of future we want to build, and navigate a path forward that balances the immense promise of CRISPR technology with the ethical imperatives that guide us.

As we delve deeper into the genetic revolution, we are not just editing genes; we're also editing our future. In the next chapter, we will look at the broader societal implications of CRISPR technology, considering how it might change our world and what that means for us all.

The Enhancement Debate: From Eliminating Diseases to Designer Babies

The journey of gene editing, from an arcane biological mechanism to a revolutionary medical technology, has been accompanied by a series of ethical dilemmas, each more challenging than the last. One of the most contentious of these dilemmas involves the distinction between the elimination of disease and the enhancement of human capabilities. Where is the line? And who gets to draw it?

The Medical Mandate

Before we delve into the ethical complexities of enhancement, it's worth reminding ourselves of the primary reason CRISPR-Cas9 has garnered so much interest and excitement: its potential to treat and even cure a host of diseases. We've long dreamed of a world without illness, and gene editing offers a tantalizing promise to move us closer to that goal.

There are over 6,000 known genetic disorders that are caused by a single mutation in the human genome. In theory, CRISPR could rectify these errors, potentially curing diseases that have been a source of human suffering for centuries. From cystic fibrosis to muscular dystrophy to sickle cell anemia, the list is long and the potential for relief profound.

The Enhancement Question

But it's not just disease eradication that has people buzzing. The same technology that can rid us of maladies can also be used to enhance human characteristics and abilities. This is where the ethical terrain becomes much more difficult to navigate.

Consider this: if it's ethically acceptable to use gene editing to enhance a crop or make a lab mouse run faster for the sake of scientific research, is it also ethically acceptable to enhance a human being? If we can make ourselves immune to HIV, why not also enhance our memory, increase our strength, or tweak our appearance?

This shift from therapy to enhancement forms the crux of the debate. Enhancement suggests moving beyond the baseline of 'normal' human functioning to create something better, stronger, or more capable. It sounds like the stuff of science fiction, but with the advent of CRISPR, it's become a genuine scientific and ethical question.

The Pros and Cons of Enhancement

On one side of the argument are those who see human enhancement as the natural next step in our evolution. Proponents argue that if we have the power to improve ourselves, to make life easier, more productive, and happier, why wouldn't we use it? After all, we routinely use technology to enhance our lives - from eyeglasses to improve our vision, to educational tools to expand our knowledge. How is genetic enhancement fundamentally different?

Opponents, however, raise a multitude of concerns. There are worries about safety - we are still learning about our incredibly complex genome, and there's much we don't understand. Unintended consequences could be serious and irreversible.

There are fears about inequality - will enhancements be available to all, or will they become another advantage for the wealthy, creating a genetic divide between the haves and have-nots?

Then there are philosophical and moral objections. Some argue that seeking to enhance ourselves is an expression of hubris, a dangerous overreach that risks undermining our appreciation of the natural human condition, with all its imperfections and variability.

The Slippery Slope to Designer Babies

The debate over enhancement is particularly heated when it comes to germline modifications. The term 'designer babies' has been coined to describe a future where parents could potentially pick and choose the traits of their children, from their intelligence to their athletic ability to their physical appearance.

Critics argue that we should not have the right to manipulate the genetic traits of future generations, that it reduces children to commodities, and that it opens up a Pandora's box of ethical and societal issues. There's also the question of where it would end. If we start with health enhancements, will we inevitably move onto enhancements for beauty, intelligence, or longevity? What will this mean for societal norms and expectations, and how will it change the human experience?

On the other hand, proponents of germline modification for enhancement purposes argue that parents have always sought to give their children every advantage, and gene editing is no different. Some even go so far as to suggest that if we have the ability to enhance our children and don't use it, we might be considered negligent.

A Matter of Governance

Decisions about how to use CRISPR technology cannot be left to scientists alone. Society as a whole has a stake in these discussions. Therefore, global consensus and regulations are needed to guide the usage of gene editing technology.

Regulation is challenging, given the global nature of science and the variation in ethical perspectives across different cultures and societies. But, to prevent a free-for-all scenario, some kind of agreement is necessary. This regulation must balance scientific progress and medical advances with the ethical, social, and legal issues raised by gene editing.

A Complex Conversation

The debate over genetic modification, particularly as it applies to human enhancement, is complex, multi-faceted, and deeply personal. It raises questions about identity, equity, disability, and the very definition of what it means to be human. There are no easy answers.

However, it is a conversation we must have, as a society and as a species. We must decide how we will use this powerful tool that we have developed. The decisions we make will shape the future in profound ways.

The next chapter of this journey will delve deeper into the societal implications of gene editing technology, considering the broader changes it might bring about and the challenges those changes might pose.

As we step into this new era, we have a responsibility to navigate it thoughtfully, conscientiously, and inclusively. After all, the story of CRISPR is not just about changing our DNA; it's about defining our values.

Chapter 4: Case Studies: CRISPR In Action

Saving Lives: Success Stories in Treating Genetic Diseases

The Revolution Begins in the Lab

As we've seen in previous chapters, the promise of CRISPR-Cas9 as a transformative tool in biology is enormous. But it is perhaps in the field of human health that the power of this technology truly comes to life. Around the globe, researchers are employing CRISPR to confront some of the most challenging diseases of our time. In this chapter, we'll delve into a series of compelling case studies where CRISPR has already begun to deliver on its promise, and examine how the technology may continue to evolve.

A Lifesaver for "Bubble Boy Disease"

One of the earliest and most dramatic examples of CRISPR's potential in medicine comes from a condition called Severe Combined Immunodeficiency (SCID). This is a rare genetic disorder often referred to as "Bubble Boy Disease" because those who have it lack a functioning immune system and must live in a sterile environment to survive.

In 2020, scientists reported a potential cure for this disease using CRISPR. The team, led by scientists from St. Jude Children's Research Hospital, used the technology to correct a mutation in the gene IL2RG, the cause of a form of SCID. In a clinical trial, eight infants with the disorder received the experimental treatment. The results were stunning: All eight began to develop functioning immune systems.

The significance of this work cannot be overstated. It provided real-world evidence that CRISPR could, in fact, be used to cure a devastating genetic disease in humans. Moreover, it highlighted the potential for the technology to be used in a wide variety of other similar conditions, heralding a new era in medicine.

Sickle Cell Disease: From Pain to Potential Cure

Next, let's turn our attention to Sickle Cell Disease (SCD), a devastating genetic condition that affects millions worldwide. This inherited disease causes red blood cells to become misshapen, leading to severe pain, organ damage, and often, shortened life span.

CRISPR is at the heart of promising new therapies for SCD. In 2019, Victoria Gray, a woman from Mississippi, became the first person in the United States to be publicly identified as being involved in a CRISPR trial for SCD. Her treatment, which involved editing her stem cells to boost the production of a type of hemoglobin, was a success. Just a year after receiving the treatment, she reported feeling stronger and free from the crippling pain that had defined her life.

This watershed moment brought hope not just for those living with SCD, but also for millions of others living with genetic diseases that could be targeted by similar therapies.

Huntington's Disease: Slowing the Inevitable

Huntington's disease, a genetic disorder that leads to progressive brain damage, has long been considered untreatable. But with CRISPR, that might change.

Scientists have used CRISPR to silence the Huntington's gene in mice, slowing the progression of the disease. While this work is still in its early stages, it offers a tantalizing glimpse into what might be possible for human patients in the future.

Leber Congenital Amaurosis: A Ray of Light

Leber Congenital Amaurosis (LCA) is a rare genetic disorder that leads to severe vision loss or blindness at birth or during the first year of life. In a breakthrough study, scientists employed CRISPR to modify the defective gene associated with this disorder, called CEP290. In late 2022, the first patient to be treated with CRISPR for inherited blindness, a nearly blind 34-year-old woman, reported substantial improvements in her sight.

As the patient shared her experiences of newly perceived brightness and discernible shapes, it offered a heartening proof-of-concept for the application of CRISPR in genetic eye disorders. Beyond LCA, the approach opened avenues to address a myriad of other inherited retinal diseases, making the idea of 'genetic sight restoration' less of a science fiction and more of an impending reality.

Cystic Fibrosis: Breathing Easier

Cystic Fibrosis (CF) is a genetic disorder that creates a buildup of thick mucus in various organs, predominantly the lungs and digestive system, affecting breathing and digestion. CRISPR has been investigated as a promising tool to correct the defective CFTR gene causing CF.

Although direct application of CRISPR therapy for CF in patients is yet to materialize, substantial strides have been made in the laboratory. Using organoids – lab-grown miniaturized and simplified versions of an organ – derived from CF patients, researchers have demonstrated successful gene correction that restored CFTR function.

These advancements in the lab provide a basis for the development of therapies that, in the future, could alleviate the symptoms and improve the quality of life for individuals living with CF.

Duchenne Muscular Dystrophy: Strength in Science

Duchenne Muscular Dystrophy (DMD) is a debilitating disease that causes progressive muscle weakness and loss. At the crux of this disorder lies a mutation in the DMD gene that results in an absence of the protein dystrophin, which is vital for muscle strength and stability.

With CRISPR's precise cutting mechanism, scientists have aimed to remove the faulty exon (a part of a gene) causing DMD, which can potentially allow the remaining exons to stitch together and form a functional – albeit shorter – dystrophin protein. Initial studies in mice and dogs have been promising, paving the way towards potential human trials.

These case studies – from blindness to lung disease to muscular dystrophy – highlight the impressive breadth of CRISPR's applicability. While some of these applications are still in the laboratory phase, others have crossed into the realm of clinical trials, bringing tangible improvements to patients' lives.

As we delve deeper into the CRISPR era, it's important to remember that every triumph also serves as a source of knowledge, helping us fine-tune the technology, making it safer and more efficient. Each case study represents a stepping stone towards a future where genetic disorders could become a thing of the past.

Agricultural Breakthroughs: High-Yield, Disease-Resistant Crops

In a world marked by increasing demands for food, strains on our agricultural resources, and the escalating threats of climate change, the application of CRISPR technology in the agricultural field represents a beacon of hope. Through precision editing of plant genomes, we can engineer crops to be more resistant to diseases and pests, boost their yield, and help them survive in extreme weather conditions, bringing a new revolution to our fields and food supply chain.

One of the first crops to bear the fruits of CRISPR's promise is rice, an essential grain that provides sustenance to billions. Scientists, using CRISPR, have successfully created a variant of rice resistant to bacterial blight, a major disease that annually claims up to 20% of the global rice crop. By disabling a gene that codes for a protein exploited by the bacteria to infect the plant, scientists turned the tables on the pathogen, blocking its invasive strategies and keeping the crop healthy.

Moving beyond disease resistance, scientists have also been tapping into CRISPR's potential to boost crop yield. For example, by editing genes related to how plants govern their energy use, researchers have developed strains of rice and potatoes with dramatically increased yields. Notably, this wasn't about making the plants bigger but rather more efficient, offering a blueprint for how we might sustainably feed an ever-growing global population.

Equally critical in today's rapidly changing climate is the development of crops that can withstand extreme weather conditions. In parts of the world where fresh water is scarce, crops traditionally dependent on abundant water supplies, such as rice, pose a challenge. However, with CRISPR, scientists have made strides in developing strains of rice capable of growing in high-salinity conditions, opening the door to their cultivation in areas previously deemed unsuitable.

Beyond staple crops like rice, CRISPR is also finding applications in fruits and vegetables. Take the case of tomatoes, where CRISPR has been used to create a variant that can resist Fusarium oxysporum, a devastating fungus that causes wilt and significant crop loss. In another instance, scientists have used CRISPR to tweak the genes of white button mushrooms to resist browning, thus extending their shelf life.

But it's not just about creating super crops; CRISPR could also play a crucial role in rescuing threatened species. The American chestnut tree, once a dominant species in North American forests, has been almost wiped out by a fungal disease. By using CRISPR to introduce a gene for fungus resistance from wheat into the American chestnut tree genome, researchers are looking to revive this near-extinct species.

Importantly, the use of CRISPR in agriculture doesn't come without its share of debates. Key among them is the question of regulation and public acceptance. Some worry about the potential ecological impacts of introducing genetically edited crops into the environment. While these concerns are valid, proponents argue that the risk can be managed through thorough testing and monitoring, and by keeping the public informed and engaged.

The implications of CRISPR's potential in agriculture are profound. By enabling us to create crops that are disease-resistant, high-yielding, and able to thrive in harsh climates, this technology offers a promising path to addressing food security and sustainability. However, as we continue to explore and apply these advances, it is crucial that we also consider the ethical, societal, and ecological implications. The future of agriculture with CRISPR looks promising, but it's a future that must be navigated with care and responsibility.

Victoria Gray

The next case study takes us to the realm of human health and the ground-breaking work done with CRISPR technology in curing inherited genetic diseases. Take the instance of sickle cell disease, a condition that affects millions worldwide, predominantly those of African descent. This genetic ailment, characterized by misshapen red blood cells that can block blood flow and cause severe pain, organ damage, and even early death, had remained without a viable cure until CRISPR entered the scene.

In 2021, Victoria Gray, a woman from Mississippi, became one of the first patients with sickle cell disease to be treated with CRISPR. The approach involved extracting her bone marrow cells, editing them with CRISPR to produce healthy red blood cells, and reintroducing them into her body. The result was a resounding success: a year after the treatment, her symptoms were gone, and she was living a life free of sickle cell disease.

Moreover, the same method is being applied to treat another blood disorder, beta-thalassemia. Early reports from clinical trials have shown promising results, and the successes seen here have given hope to millions living with these debilitating conditions.

Another promising application of CRISPR in healthcare is in the realm of cancer treatment. The year 2020 saw the first CRISPR-based clinical trial for cancer at the University of Pennsylvania, where patients with refractory cancer received CRISPR-edited T cells. The aim was to equip the patient's immune system to better combat cancer, and although the study was primarily to establish the treatment's safety, it was encouraging to see some patients respond positively.

These are just a few examples of how CRISPR is making waves in medicine. Other areas, like inherited blindness, muscular dystrophy, and even HIV, are also being explored, promising an era where previously untreatable genetic conditions can be effectively managed, if not outright cured.

However, as with the agricultural applications of CRISPR, its use in healthcare also raises significant ethical and regulatory issues. For one, editing the human genome—particularly making changes that can be inherited (germline modifications)—opens a Pandora's box of ethical questions. Further, there's also the question of accessibility: as with many breakthrough treatments, the cost of CRISPR-based therapies is currently high, raising concerns about who will be able to benefit from these advances.

In summary, the case studies in this chapter demonstrate the profound potential of CRISPR, from feeding the world sustainably to curing previously intractable diseases. But they also highlight the critical importance of carefully navigating the ethical and societal implications of this technology.

As we move forward, ensuring responsible, inclusive, and ethical use of CRISPR will be crucial in realizing its full potential for humanity's benefit. In the next chapter, we will discuss these ethical and societal implications in greater detail, providing a comprehensive understanding of the broader context within which this revolutionary technology operates.

The Dangers: The Case of He Jiankui and Designer Babies

The power of CRISPR and the prospect of genome editing bring not only immense promise but also significant perils. One such peril that stunned the global scientific community was the case of He Jiankui, a Chinese scientist who undertook a rogue experiment that resulted in the world's first genetically edited babies. His case serves as a cautionary tale about the abuse of CRISPR and underlines the urgent need for stringent regulations and ethical guidelines surrounding its use.

In 2018, He Jiankui announced to the world that he had used CRISPR to edit human embryos, which were then implanted into a woman who gave birth to twin girls, known by their pseudonyms, Lulu and Nana. He's stated goal was to make the babies resistant to HIV by disabling a gene called CCR5, which the virus uses as a pathway to invade human cells.

This announcement was met with a wave of shock and condemnation from the scientific community worldwide. Many scientists considered his actions to be grossly unethical and a blatant disregard for the widely accepted scientific consensus against modifying human embryos destined for birth.

The reasons behind this global outrage are multifold. First, the experiment lacked transparency. The details of the procedure were not published in a peer-reviewed journal, a standard process that allows other scientists to evaluate the methodology and validity of the results. Second, the experiment was not medically necessary. While HIV is a serious health issue, many other treatments and preventive measures are available. Genome editing was far from the last resort.

Furthermore, the potential long-term risks of the edits made by He Jiankui are hard to predict. While CRISPR is a precise tool, it isn't perfect. There can be off-target effects, where genes other than the intended ones are altered, and these could have unforeseen and potentially harmful consequences. Additionally, the deletion of the CCR5 gene, while conferring resistance to HIV, has been linked to an increased vulnerability to other viruses such as West Nile and influenza.

Perhaps the most unsettling aspect of He Jiankui's experiment was that the genetic modifications he introduced into Lulu and Nana's genomes are germline changes, meaning they could be passed down to future generations. The long-term implications of these edits on the human gene pool are unknown and could potentially introduce new diseases or health problems.

He Jiankui's actions, in essence, opened a Pandora's Box of genetic modification, ushering in a new era of "designer babies," where the genetic attributes of children are predetermined by their parents or scientists. This brings to the fore a whole new set of ethical questions. How do we determine what traits are desirable? How do we prevent the creation of a socio-economic divide between those who can afford gene editing and those who cannot? Could we unintentionally create a new form of eugenics?

These questions underscore the vital importance of ethical considerations and regulatory oversight in the use of CRISPR technology. He Jiankui's actions led to a prison sentence in China and have spurred scientists and policymakers around the world to push for clearer and stricter guidelines for the use of gene-editing technologies in humans.

In the aftermath of this incident, the World Health Organization established a global registry for human genome editing research to promote transparency and responsibility. Various scientific bodies and countries are also working on establishing their guidelines, emphasizing ethical considerations, expert consultations, and public dialogue.

The case of He Jiankui serves as a potent reminder of the potential misuse of CRISPR technology. It also highlights the pressing need for the global community to establish a robust ethical and regulatory framework to prevent such abuses and ensure that genome editing is used responsibly, ethically, and to the benefit of all humanity.

As we move forward in the genetic age, learning from such instances and taking proactive measures will be crucial. We stand on the brink of an era where our ability to manipulate the very building blocks of life can lead to profound breakthroughs. From curing genetic diseases to creating more sustainable and resilient crops, the possibilities are boundless. But this same power can also lead to troubling consequences if misused or placed in the wrong hands, as the case of He Jiankui demonstrates.

The critical task before us is to navigate this brave new world with responsibility, ethical consideration, and foresight. We must establish robust oversight systems, informed by ongoing dialogue between scientists, ethicists, policymakers, and the public. The purpose of these systems should not be to stifle innovation but to ensure it's used in ways that align with our shared values and principles, and for the benefit of all humanity.

Moreover, as we unlock the potential of CRISPR, we need to ensure that it does not exacerbate existing inequalities or lead to new ones. Access to its benefits should not be limited to those who can afford it. Instead, we must strive to create a world where everyone, irrespective of their economic status, can benefit from the advances in gene editing technology.

Furthermore, we need to invest in public education and transparency around genome editing technologies. The public should be informed about the possibilities and risks associated with these technologies and be engaged in discussions about their use. This would allow a more informed and inclusive debate about the ethical boundarics we want to set for gene editing.

The case of He Jiankui serves as a stark reminder of the ethical complexities and potential dangers of CRISPR technology. However, it should not deter us from harnessing its potential. Instead, it should motivate us to engage in an ongoing dialogue, create robust regulatory frameworks, and make thoughtful decisions that balance the promise of CRISPR with the ethical, societal, and health considerations it raises.

We are at a crossroads in our journey with CRISPR. The decisions we make now will not only shape the future of this technology but also the future of our species. As we continue to delve deeper into the CRISPR universe, we should do so with care, curiosity, and a firm commitment to uphold the highest ethical standards.

Chapter 5: Societal Implications: Access, Equity and Governance

Accessibility and Affordability: Who Gets to Edit?

Let's embark on an exploration of the contentious issue of accessibility and affordability in the realm of gene editing. As with most advancements in science and technology, CRISPR has the potential to create new disparities while mitigating existing ones.

The unveiling of the CRISPR technology promised a democratization of genetic engineering, making it more affordable and accessible to laboratories across the world. No longer was it confined to the high-tech labs with vast resources; suddenly, the power to edit genes was available to a much wider scientific community. This has led to an explosion of research and experimentation, accelerating our understanding of genetics and opening up new avenues for disease treatment and prevention.

However, as we begin to move from laboratories to clinical applications, we face a new question: Who gets to reap the benefits of this revolutionary technology? Who gets to edit?

This is a deeply complex issue, interweaving threads of economics, ethics, and social justice. For all its promise, CRISPR-based therapies could be expensive, potentially limiting access to wealthy individuals and countries that can afford it. This is not a new scenario. In fact, we have already seen it unfold with other innovative treatments, such as the Hepatitis C drug Sovaldi, or CAR-T therapies for certain types of cancer, which come with price tags in the tens, or even hundreds, of thousands of dollars.

Should gene editing follow this trend, we risk creating a genetic divide, where the rich have the means to eliminate genetic diseases, enhance their abilities or even, potentially, extend their lifespans, while the poor continue to suffer. This could further exacerbate social and economic disparities that are already deep-rooted in our societies.

Then there's the issue of global disparity. Currently, much of the research and development in CRISPR technology is concentrated in a handful of countries, primarily in the Western world. Many countries in the Global South may lack the resources, infrastructure, or expertise to develop or implement CRISPR-based therapies, potentially widening the health and technological gap between the Global North and South.

On the other hand, there is hope. The potential of CRISPR to alleviate human suffering is so vast that it has stimulated discussions about how to ensure its equitable distribution.

Some have suggested a tiered pricing system, where the cost of treatment is scaled according to the country's income level, similar to strategies used for antiretroviral therapies for HIV.

Others propose a global fund, much like the Global Fund to fight AIDS, Tuberculosis, and Malaria, that would help subsidize the cost of CRISPR treatments in low-income countries. A universal genome editing fund could help level the playing field and ensure that CRISPR's benefits are not confined to the world's wealthiest individuals or countries.

The final, and perhaps most important, piece of this puzzle is governance. It's clear that ensuring equitable access to CRISPR will require robust, globally-coordinated oversight, and a commitment to principles of social justice and equality. This includes setting international standards for price and access, monitoring adherence to these standards, and investing in capacity building in lower-income countries to enable them to harness CRISPR's power.

Overall, the question of "Who gets to edit?" will have far-reaching implications for social equity in the age of CRISPR. It's a question that demands our attention, our empathy, and our best creative problem-solving. As we stand at this genetic crossroads, let's commit to ensuring that the path we take leads to a world where everyone can benefit from the incredible potential of CRISPR.

Governing CRISPR: Regulatory Landscapes Around the World

In a world where the power of gene editing is becoming increasingly accessible, the question of how we govern and regulate the use of technologies such as CRISPR-Cas9 becomes increasingly critical. Different nations, with their diverse cultural, social, and political landscapes, are choosing different paths in regulating CRISPR, creating a patchwork of policies and guidelines that reflects both the opportunities and the challenges posed by this technology.

Starting in the West, the United States has approached the regulation of CRISPR with what might be described as cautious optimism. The Food and Drug Administration (FDA) considers genetically edited organisms to be drugs and therefore subject to existing regulation. The National Institutes of Health (NIH) has also been careful in funding CRISPR research, particularly when it comes to potential applications in humans, steering clear of germline modifications due to ethical concerns.

The European Union, by contrast, has adopted a more precautionary approach. The Court of Justice of the European Union ruled in 2018 that gene-edited organisms should be classified as genetically modified organisms (GMOs), making them subject to stringent regulations. This decision has faced criticism from some scientists, who argue that it could stifle innovation and impede progress in sectors like agriculture, where CRISPR could be used to produce more resilient and sustainable crops.

China's regulatory landscape has been notably less restrictive, which has both facilitated groundbreaking research and sparked controversy. The world's first gene-edited babies, created by scientist He Jiankui in defiance of international ethical norms, were born in China, leading to international outcry and highlighting the dangers of insufficient regulation.

The disparities in these regulatory landscapes have significant implications. On one hand, varying regulations can foster a diversity of research environments, promoting innovative experimentation in some regions, while encouraging caution and rigorous scrutiny in others. On the other hand, the lack of international consensus on the governance of CRISPR presents significant challenges. Divergent regulations can result in 'CRISPR tourism', where individuals or companies move their operations to countries with less stringent regulations. This not only poses ethical dilemmas, but could also lead to safety issues and abuses of the technology.

Many experts believe that the solution lies in fostering international cooperation and harmonizing regulations. Several bodies, such as the World Health Organization, have started to work on establishing global standards for human genome editing. These efforts aim to create a regulatory floor, a set of minimum standards that every country should meet, while leaving room for individual nations to impose stricter regulations if they choose.

The regulatory landscapes governing CRISPR around the world are as varied as the potential applications of the technology itself. Balancing the need for innovation with ethical and safety considerations is a delicate task. As we continue to navigate this new frontier, cooperation, transparency, and a commitment to shared ethical principles will be essential in shaping regulations that guide us towards the responsible use of this powerful technology.

The governance of CRISPR is not just about controlling a technology; it is about shaping our shared genetic future. Let us ensure that this future is characterized by equity, caution, and respect for the immense power we wield.

The Role of Public Perception and Media in Shaping the CRISPR Debate

The emergence of CRISPR-Cas9 technology, with its profound implications for the future of humanity, has catalyzed a great deal of public interest and debate. As we grapple with the ramifications of this genetic revolution, one of the most significant factors shaping the discourse is the public's perception of the technology, which in turn, is heavily influenced by media portrayals.

In recent years, CRISPR has received extensive coverage across diverse media platforms, from newspapers and magazines to social media and documentaries. The way this technology is presented in these different outlets plays a pivotal role in shaping the public's understanding and attitudes towards gene editing.

The framing of CRISPR in media narratives can range from the sensational to the sober. On one hand, there are headlines touting the arrival of a brave new world, replete with designer babies and potential superhumans. These tend to amplify the more sensational aspects of gene editing, sometimes oversimplifying the science and overlooking the significant technical and ethical challenges that remain. Such narratives can generate both hype and fear, fostering unrealistic expectations or alarmist reactions.

On the other hand, more nuanced accounts delve into the complexities and potential of CRISPR, explaining the science in layman's terms and balancing the potential benefits with the ethical concerns. These help educate the public, grounding the discourse in fact rather than speculation, and fostering informed debates.

In both cases, the media plays an instrumental role in influencing public perception and, by extension, the public's role in the CRISPR debate.

Public perception of CRISPR, in turn, has a significant impact on policy-making and regulation. Lawmakers are responsive to the concerns and expectations of their constituents, and public opinion can shape the political will for funding research, crafting regulation, or promoting certain applications of the technology. This influence can be seen in various countries' approaches to regulating CRISPR, from the EU's cautious restrictions to China's more permissive stance.

However, ensuring that public opinion is informed by accurate understanding of the technology is a significant challenge. Public understanding of genetic science is generally low, and CRISPR's complexities can be difficult to communicate effectively. This underscores the importance of science communication, both in the media and in educational settings.

Science communication can foster a deeper understanding of the technology and its implications, equipping the public to participate more effectively in the societal debate on CRISPR. It can also help to dispel misconceptions and alarmist reactions, by providing accurate information and a nuanced perspective on the potential and limits of gene editing.

The media's role in shaping public perception of CRISPR, and the influence of this perception on the societal discourse and policy decisions surrounding the technology, cannot be overstated.

As we navigate the future of this groundbreaking technology, it is crucial that we strive for clear, accurate, and balanced representation in the media, and foster an informed and engaged public.

The decisions we make about the use and regulation of CRISPR are decisions about our collective future. Ensuring that these decisions are informed by a deep and nuanced understanding of the technology and its implications is essential as we stand at the crossroads of this genetic revolution.

Chapter 6: A Brave New World: Hypothetical Scenarios and Their Ethical Implications

Scenario Analysis: Utopian and Dystopian Visions

The possibilities brought about by the advent of CRISPR gene-editing technology are not simply matters of science. They open doors to a labyrinth of moral, ethical, and societal questions that we must strive to answer as we stand on the cusp of this genetic revolution.

What lies on the other side of this technological threshold are myriad scenarios, some utopian, some dystopian, each encapsulating the hopes, fears, dreams, and concerns inherent to the human condition. In this chapter, we delve into some of these possible futures, unravelling their implications, and urging us all to carefully consider the kind of world we wish to create.

Let's first imagine a utopian vision, where the power of CRISPR is harnessed optimally and ethically. Diseases once thought incurable—cystic fibrosis, Huntington's disease, Duchenne muscular dystrophy—are eradicated from the genetic code. Malaria and Zika are distant memories, as genetically modified mosquitoes no longer carry these plagues. The world's food crisis is eased as high-yield, disease-resistant crops flourish in fields across the globe. Even climate change is addressed, with genetically modified organisms efficiently capturing carbon dioxide, curbing our greenhouse gas emissions.

In this world, the technology is accessible to all, governed by clear, stringent international regulations that prioritize the public good. There is no room for designer babies; genetic modification is solely utilized to combat disease and improve quality of life.

But let's also cast our gaze on the potential dystopian future, where unchecked gene editing might lead us. In this world, the promise of CRISPR is eclipsed by rampant misuse. Genetic modifications are a luxury of the rich, leading to an unprecedented socio-genetic divide. The wealthy enhance their offspring with desirable traits, from intelligence to physical prowess, leading to a new form of eugenics.

Worse still, rogue states or unscrupulous entities might create genetically modified bioweapons, unleashing unpredictable and potentially catastrophic consequences. The risks of off-target effects and unforeseen consequences of genetic modification are ignored, leading to the inadvertent introduction of new diseases.

These scenarios, while dramatically opposed, serve to underscore the significant ethical implications of CRISPR technology. They remind us that while science provides us with increasingly powerful tools, how we use them is a reflection of our values, our ethics, and ultimately, our humanity.

As we navigate the path ahead, we must be mindful of the risks and potential pitfalls that come with the power to manipulate life's fundamental code. In our pursuit of progress, we must not lose sight of the inherent dignity and worth of all individuals, regardless of their genetic makeup.

By actively engaging in these debates, and by grappling with these difficult questions, we can help shape a future where gene-editing technologies like CRISPR are used to better our world, and not to divide or harm it. It is a future that we all have a stake in, and one that will require wisdom, foresight, and an unwavering commitment to the common good.

In the words of Carl Sagan, we are "a way for the cosmos to know itself". With the advent of CRISPR, we have been given the tools to know ourselves at an even deeper level, to manipulate the very essence of what makes us who we are. As we stand at the crossroads of this brave new world, we are compelled to ask ourselves - what will we do with this knowledge? How will we, as a society, choose to wield this extraordinary power? As we ponder these questions, we must strive to guide this genetic revolution with wisdom, compassion, and a profound respect for all life.

Ethical Analysis: Applying Ethical Theories to Scenarios

Embarking on the exploration of ethics in the context of CRISPR and gene-editing technology calls upon us to engage with established ethical theories and apply them to the scenarios presented by this cutting-edge science. Ethics, in its broadest sense, serves as a guide for moral decision-making. It offers a platform for the evaluation of what is right, just, and fair.

The two primary ethical theories we'll consider in this chapter are consequentialism and deontology, as they are often found at opposite ends of the ethical spectrum. Consequentialism judges an action based on its consequences, positing that the most ethical action is the one that results in the most good. Conversely, deontology holds that the morality of an action is inherent in the action itself, irrespective of its outcomes.

Let's examine these theories through the lens of gene editing, and dive into the potential scenarios that might play out in both our utopian and dystopian visions.

Firstly, consequentialism. If we apply this ethical framework to our utopian scenario, where CRISPR is used to eradicate disease, increase food production, and combat climate change, the assessment is quite clear. The outcomes are overwhelmingly positive, serving the greater good, and therefore, the application of gene-editing technology in this manner is ethically sound.

But what if we take our dystopian vision, where genetic enhancements are exclusive to the wealthy and lead to socio-genetic divide? Here, consequentialism might present a divided verdict. While the wealthy minority benefits, creating a more intelligent, healthier, and potentially happier subgroup, it leaves the majority at a significant disadvantage, leading to increased social inequality. Therefore, the ethical nature of this application becomes questionable under consequentialism, as it does not result in the most good for the most people.

Now, let's look at deontology, which emphasizes duty, rules, and the inherent morality of actions. In our utopian scenario, deontology might object to the very act of gene editing, arguing that it is morally incorrect to tamper with the natural genetic order, regardless of the positive outcomes. Deontologists might argue that we have a duty to respect the natural genetic process and uphold the intrinsic dignity of all human life, unaltered and unedited.

In our dystopian scenario, deontology might present an even stronger objection. Here, gene editing is not only breaching the natural order but also creating a new form of social discrimination based on genetic makeup. This application is not just morally incorrect due to the act of gene editing itself, but also due to its violation of the principle of equal respect and dignity for all.

However, it is essential to note that these ethical theories do not provide absolute answers. They are guiding frameworks, intended to inform and shape our moral judgments. In real-world situations, it's often necessary to balance these different ethical perspectives.

Given the potential impact of CRISPR and gene-editing technology, it is crucial to engage in these ethical discussions, examining our actions from various moral standpoints. This reflection will help ensure that as we move forward, we do so with careful consideration of not just the scientific and medical implications of our actions, but also the moral and ethical ones.

The future of CRISPR is a journey into uncharted territory, filled with promise and perils. It is up to us, as a global community, to navigate this course in a manner that respects our shared ethical values and promotes the greater good. The ethical analysis applied here is a tool to guide us, reminding us that while we wield the power to edit life's fundamental code, we have a responsibility to do so with great care and respect for all life.

Conclusion: CRISPR and the Future of Humanity

Recap of the CRISPR Journey

As we draw to the close of our journey through the multifaceted landscape of CRISPR, it is appropriate to take a step back and absorb the profound magnitude of what we've explored. The CRISPR-Cas9 technology, with its potential to manipulate life at its most fundamental level, has irreversibly changed the course of biological sciences, raising questions that extend far beyond the confines of labs and research institutions.

Our journey began with an exploration of the science underlying CRISPR. We dove into the world of genetic codes and molecular scissors, understanding how the technology has revolutionized gene editing. This is a technology that allows us to manipulate DNA, the essence of life, with an ease and precision hitherto unimaginable. It offers a promise to potentially cure genetic diseases, bolster food security, and even combat climate change.

We delved into the ethical dimensions of this technology, treading the line between disease treatment and genetic enhancement. Here, we explored how the application of gene editing raises crucial questions about morality, the sanctity of life, and what it means to be human. It's a technology that prompts us to examine our values and forces us to reckon with the paradox of progress: with every leap forward, we must be mindful of the ethical dilemmas and moral quandaries that arise.

We touched upon the dramatic implications of germline modifications, where changes made are not limited to the individual but echo into future generations. In this context, the importance of cautious progression is heightened, as we must consider not just the rights and health of the current generation, but of those yet unborn.

Our exploration continued into the realm of societal implications, delving into questions of access, affordability, and governance of CRISPR technology. We grappled with the challenges of ensuring equitable access, and the potential for socio-genetic stratification. The need for comprehensive, globally aligned regulations was underscored, as we assessed the varying regulatory landscapes around the world.

Lastly, we ventured into the world of speculative scenarios, exploring both utopian and dystopian visions brought about by gene editing. We pondered over how gene editing could lead to enhanced humans with improved cognitive abilities and lifespan. Conversely, we also considered dystopian scenarios, where genetic enhancements could lead to societal divides, and the risk of rogue uses.

As we stand at this precipice of possibility, we cannot overstate the importance of robust, informed, and inclusive discourse. While scientists and researchers carry forward their groundbreaking work, it is incumbent upon all of us—scientists, ethicists, lawmakers, and the public—to engage in the discussions and decisions that shape our genetic future. The road forward is uncharted and the challenges are complex, but with thoughtful navigation, the potential for good is immense.

Our CRISPR journey is far from over. As we continue to push the boundaries of what we can achieve, we must also constantly redefine what we should achieve. This technology holds the potential to shape the future of humanity, and it's a future that we must craft together, with an acute awareness of the ethical, societal, and scientific dimensions at play.

Critical Considerations for the Path Forward

As we stand at the crossroads of a brave new world, with CRISPR technology in our hands, it is of the utmost importance to consider carefully the path we choose to tread. Armed with the power to rewrite the genetic code of life itself, our decisions today will echo into the eternity of humanity's future. We cannot afford to take lightly the immense potential and perils that this technology presents. In this chapter, we will focus on the crucial considerations that must shape our approach to the future of CRISPR and gene editing technologies.

First and foremost, the need for comprehensive and dynamic regulatory frameworks cannot be overstated. The case of He Jiankui, the Chinese scientist who created the world's first genetically edited babies, is a stark reminder of the risks of unregulated use of CRISPR technology. The international scientific community condemned his actions, but the fact that he could proceed with such a profoundly consequential experiment in the absence of clear-cut regulations is deeply troubling.

Different nations have disparate regulations, reflecting the cultural, ethical, and societal views of their populations. Some, like the UK, have a relatively permissive approach to embryo research, while others, like Germany, have stringent laws reflecting historical sensitivities. Harmonizing these regulatory landscapes will be a challenge, but it is an endeavor we must undertake. Global scientific cooperation and multilateral dialogues will be critical in this regard.

Secondly, the importance of ethical considerations in guiding our use of CRISPR technology is paramount. While CRISPR holds potential for treating debilitating genetic diseases, it also opens the door to genetic enhancements, leading to 'designer babies.' Questions surrounding the morality of such enhancements, and the potential for creating a socio-genetic divide, must be addressed. We must ask ourselves: just because we can, should we?

These discussions should not be confined to academic circles or legislative chambers; they need to involve the public at large. Public engagement in the discourse around CRISPR technology is essential to ensure that the path we take aligns with societal values and moral principles. Transparency in scientific endeavors, education to raise public understanding, and open dialogue are vital in this regard.

The media plays an indispensable role in shaping public perceptions and driving discourse around CRISPR. Therefore, the responsibility placed on journalists and media outlets is immense. They must strive to present accurate, balanced, and accessible information on the developments in CRISPR technology, steering clear of sensationalism.

Additionally, accessibility and affordability are significant considerations. If the benefits of CRISPR are to be truly widespread, we must address the potential for socio-genetic stratification, where only the affluent can afford genetic treatments or enhancements. The democratization of CRISPR technology must be a priority, ensuring that this revolutionary technology does not widen existing societal inequalities.

A related concern is the potential misuse of CRISPR technology. The ease and low cost of CRISPR make it accessible to a wide array of actors, increasing the risk of misuse, whether it be in the form of bioterrorism or the creation of harmful genetically modified organisms. Here again, robust regulatory frameworks and international cooperation are crucial.

Lastly, the need for further research and development is critical. While CRISPR has shown immense promise, there is still much we do not understand. Concerns about off-target effects and long-term impacts need to be addressed. It's crucial to support ongoing research to refine the technology, improve its accuracy, and expand its applications.

The path forward for CRISPR technology must be navigated with a sense of collective responsibility, a commitment to ethics, and a vision for the betterment of all humanity. We must acknowledge and respect the power of the tool we hold, balancing the promise it holds with the potential perils it presents. Our journey with CRISPR is far from over, but the critical decisions we make now will define our collective destiny.

As we look to the future, we cannot underestimate the impact of continuing technological advances. The dynamic and rapid nature of CRISPR research demands constant vigilance and responsiveness from those charged with regulation and oversight. It's not just a matter of creating the right regulations now; it's about maintaining an ongoing dialogue and readiness to adapt and modify regulations as the technology evolves. This cannot be a one-time effort, but a sustained commitment.

Moreover, in the coming years, as we continue to grapple with the global challenges such as climate change, food security, and health inequities, CRISPR has the potential to be a game-changer. From creating disease-resistant crops to combat food scarcity, to treating genetic diseases and potentially curing cancer, the opportunities are truly transformative. But to leverage CRISPR's potential to solve these global issues, international cooperation is a necessity. We are in this together, and the challenges of our shared future require shared solutions.

The principle of equity, too, should be at the heart of our path forward. As we've noted, the power of gene editing carries the risk of deepening social disparities if its benefits are only accessible to the privileged few. Mitigating this requires both policy interventions to ensure broad access and affordability, and also a commitment to education and empowerment. To ensure truly equitable access, we must strive for a world where everyone understands this technology that holds such sway over our futures, where everyone can engage meaningfully in the conversations about how we use it.

As we move forward, it's crucial to remember that the real power of CRISPR doesn't lie in the protein Cas9 or the guide RNA, but in the hands of the people wielding it. This power can be used responsibly or recklessly, for the benefit of all, or to the detriment of many. Hence, the most critical consideration for the path forward is us – scientists, policymakers, journalists, ethicists, and the public.

How we handle the power of CRISPR, how we make collective decisions about its use, and how we manage its risks and benefits, is up to us. Our actions and choices will write the story of CRISPR, and indeed, the story of humanity in the age of genetic editing.

Navigating the path forward with CRISPR will undoubtedly be fraught with challenges, but also ripe with potential. The promises of CRISPR are boundless, from eradicating genetic diseases to ensuring food security, from personalizing medicine to possibly extending our lifespans. At the same time, the perils are profound, necessitating our utmost caution and respect for the power we wield. We stand at the brink of a brave new world, and the steps we take now will shape our shared future.

CRISPR is not merely a tool; it represents a symbol of human progress, a testament to our never-ending pursuit of knowledge, and an embodiment of our collective potential and perils. As we continue this journey, we must strive for a future where CRISPR is used responsibly, ethically, and for the betterment of all humanity. And as we walk this path, let us remember that every step we take is writing a new chapter in the annals of human history. Let's strive to make it one that future generations will look back upon with gratitude and pride.

The CRISPR journey continues, and it is ours to shape.

The Role of Society in Shaping the CRISPR Revolution

As we navigate the era of the CRISPR revolution, it becomes increasingly evident that society plays an integral role in shaping this journey. Every scientific discovery exists in a social context, and CRISPR is no exception. To think of CRISPR as purely a biological tool would be to overlook the myriad of social, ethical, and political considerations intertwined with its application.

There is a prevailing belief that science and society exist in separate domains. However, in reality, they are interconnected facets of our human experience, influencing and being influenced by each other. A tool as potent as CRISPR, with its profound potential to alter the course of life, blurs these boundaries even further, bringing society right to the center of the scientific conversation.

At its heart, the CRISPR revolution is about more than just the scientific potential of gene editing. It's about power - who has it, who doesn't, and how it's used. In this sense, CRISPR becomes a lens through which we can view larger societal dynamics and inequities.

One of the primary roles of society in this revolution is that of gatekeeping. Through legislative and regulatory mechanisms, society sets the boundaries for what is deemed permissible. In essence, we collectively decide the limits of the CRISPR frontier. This gatekeeping function isn't just about what we can do with CRISPR, but more fundamentally, what we should do.

The legislative arena is a battleground where various societal forces converge. Here, politicians, scientists, ethicists, and advocacy groups, all armed with their own perspectives and agendas, debate and negotiate the path forward. But laws and regulations are not static entities. They evolve, reflecting changing societal norms and attitudes. They encapsulate society's collective judgement and serve as a social contract that governs our collective exploration of the CRISPR frontier.

Public opinion and societal attitudes also play a crucial role. These can shape the trajectory of scientific research, pushing it forward or holding it back. In democratic societies, public sentiment can influence political will, which in turn impacts legislation and funding. Hence, the conversation about CRISPR cannot be confined to scientific and legislative circles. It needs to be an inclusive dialogue that invites and respects diverse perspectives. Engaging the public in the CRISPR conversation is essential not only for the democratic process but also for building trust and legitimacy.

Media, as a powerful conduit of information and shaper of public opinion, plays a pivotal role in this dialogue. It mediates the conversation between the scientific community and the public, translating the complex language of science into accessible narratives. In the

CRISPR revolution, the media's role is magnified due to the profound implications and potential misunderstandings surrounding gene editing. It holds the potential to illuminate or obfuscate, to calm fears or ignite them, to build bridges or widen chasms.

Education, too, is a cornerstone of the societal role in the CRISPR revolution. In an era where gene editing has become a reality, genetic literacy is no longer a luxury; it's a necessity. By equipping individuals with knowledge, we empower them to engage in informed conversations about CRISPR, to voice their opinions, and to make decisions that impact their lives.

Moreover, society has the responsibility to ensure equity in the CRISPR revolution. The question of who has access to CRISPR's benefits, who gets to participate in the decision-making process, and who bears the risks, are deeply entwined with societal structures and inequities. It's incumbent upon us to address these disparities and to strive for a future where CRISPR is used for the benefit of all, not just the privileged few.

Finally, the societal role extends to grappling with the profound ethical dilemmas that CRISPR presents. These are not questions that science can answer on its own. They are deeply human questions that require us to examine our values, our fears, and our hopes.

These ethical dilemmas are not just abstract intellectual exercises. They are real, tangible issues that will affect lives in profound ways. Consider, for example, the question of editing the human germline. This isn't just a scientific question; it's a societal one. It requires us to ponder on the kind of future we want, the kind of humans we aspire to be, and the legacies we wish to leave for our descendants.

Society also plays a vital role in guiding the narrative around CRISPR. As powerful as CRISPR is, it's not an inevitable force that will shape our future without our input. We, as a society, hold the reins. Our choices, our voices, and our collective will, are what will ultimately define the CRISPR revolution.

Moreover, society holds the power to ensure that the conversation around CRISPR is inclusive. By creating platforms for diverse voices to be heard, we can ensure that the decisions we make about CRISPR are representative of a broad range of perspectives, not just those of a select few. Ultimately, the role of society in shaping the CRISPR revolution is a multifaceted one. It extends from the legislative halls where laws are drafted, to classrooms where the next generation of citizens is educated, to dinner tables where family members share their hopes and fears about gene editing.

This role isn't a passive one. It demands active engagement, critical thinking, and open dialogue.

It requires us to challenge our assumptions, to listen with empathy to differing views, and to navigate complex ethical landscapes with humility and courage.

In the face of a tool as powerful and transformative as CRISPR, we, as a society, have a responsibility to guide its trajectory in a way that aligns with our shared values and vision for the future. This is no small task, but it is within our reach. The future of CRISPR, and ultimately, the future of humanity, is a shared endeavour – one that we must undertake together.

To conclude, the CRISPR revolution is as much a societal journey as it is a scientific one. And in this journey, society's role is crucial. It's the role of the gatekeeper, the educator, the ethical guide, the narrative shaper, and the advocate for equity.

By embracing this role, we can ensure that the CRISPR revolution is not just about editing genes, but also about fostering a future that reflects our collective hopes, values, and aspirations.

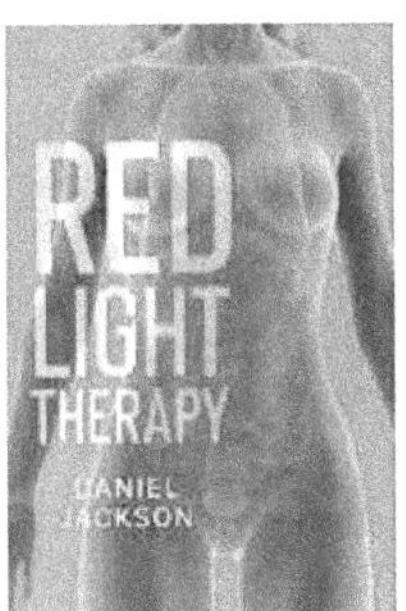

Take a look at more great books available from Rockwood Publishing

... some for FREE!

Just visit the link below:

rockwoodpublishing.co.uk

to the subject matter covered. The information included in this book has been compiled to give an overview of the subject(s) and detail some of the symptoms, treatments etc. that are available to people with this condition. It is not intended to give medical advice. For a firm diagnosis of your condition, and for a treatment plan suitable for you, you should consult your doctor or consultant. The writer of this book and the publisher are not responsible for any damages or negative consequences following any of the treatments or methods highlighted in this book. Website links are for informational purposes and should not be seen as a personal endorsement; the same applies to the products detailed in this book. The reader should also be aware that although the web links included were correct at the time of writing, they may become out of date in the future.

Disclaimers

The content contained within this book is for information and entertainment purposes only, and in no way purports to represent professional medical opinion. It should NOT be used as a substitute for expert advice, and you must consult with your designated health professional before acting upon any information contained herein or before undertaking any practice whose methodology is referred to in this book. The author is NOT a registered health professional and the text merely represents personal opinion, not medical fact. The author cannot be held responsible for the consequences of any action derived from the reading of this book, as the content is not based on diagnosis and subsequent regimen. It is the reader's responsibility to seek proper, professional medical advice from a registered health practitioner in connection with any material contained within this book.

Legal Disclaimer (part 1)

Nothing in this book should be construed as an attempt to diagnose, treat or cure. The information in this book is intended to be a community resource. The author takes no responsibility for any informational material or brochures produced using information taken from this book. The author has endeavoured to ensure that all information is correct at the time of publication. This information, however, is subject to change without notice. The author makes no warranty with regard to the accuracy of any information and will not be liable for any errors or omissions. Any liability that arises as a result of this information is hereby excluded to the fullest extent allowed by law.

This information should not be used as a substitute for seeking independent professional advice.

Legal Disclaimer (part 2)

Disclaimer and Terms of Use:

a) i. In publishing this information, the author makes no representations concerning the efficacy, appropriateness or suitability of any products or treatments. Use this information at your own risk. The compiler is not a doctor and has no medical background or training.

ii. Statements and information regarding dietary supplements, books and any products mentioned have not been evaluated by any health authority and are not intended to diagnose, treat, cure or prevent any disease or health condition.

b) In view of the possibility of human error, neither the author nor any other party involved in providing this information, warrant that the information contained therein is in every respect accurate or complete and they are not responsible nor liable for any errors or omissions that may be found or for the results obtained from the use of such information. The entire risk as to use of this information is assumed by the user.

c) You are encouraged to consult other sources and confirm the information.

d) The information you access is provided "as is". No warranty, expressed or implied, is given as to the accuracy, completeness or timeliness of any information herein, or for obtaining legal advice. To the fullest extent permissible pursuant to applicable law, neither the author nor any other parties who have been involved in the creation, preparation, printing, or delivering of this information assume responsibility for the completeness, accuracy, timeliness, errors or omissions of said information and assume no liability for any direct, incidental, consequential, indirect, or punitive damages as well as any circumstance for any complication, injuries, side effects or other medical accidents to person or property arising from or in connection with the use or reliance upon any information contained herein.

e) The author is not responsible for the contents of any linked site or any link contained in a linked site, or any changes or update to such sites. The inclusion of any link does not imply endorsement by the author. The author makes no representations or claims as to the quality, content and accuracy of the information, services, products, messages which may be provided by such resources, and specifically disclaims any warranties, including but not limited to implied or express warranties of merchantability or fitness for any particular usage, application or purpose.

f) The information provided is general in nature and is intended for educational and informational purposes only. It is not intended to replace or substitute the evaluation, judgment, diagnosis, and medical or preventative care of a physician, paediatrician, therapist and/or health care provider.

g) Any medical, nutritional, dietetic, therapeutic or other decisions, dosages, treatments or drug regimes should be made in consultation with a health care practitioner. Do not discontinue treatment or medication without first consulting your physician, clinician or therapist.

h) By reading this information, you signify your assent to these terms and conditions of use. If you do not agree to these terms and conditions of use, do not read/use this information. If any provision of these terms and conditions of use shall be determined to be unlawful, void or for any reason unenforceable, then that provision shall be deemed severable from this agreement and shall not affect the validity and enforceability of any remaining provisions.

i) The information, services, products, messages and other materials, individually and collectively, are provided with the understanding that the author is not engaged in rendering medical advice or recommendations.

j) The information and the terms of use are subject to change without notice. The material provided as is without warranty of any kind and may include inaccuracies and/or typographical errors. The author makes no representations about the suitability of this information for any purpose. The author disclaims all warranties with regard to this information, including all implied warranties, and in no event shall the author be held liable, resulting from, or in any way related to, the use of this information.

k) The unauthorized alteration of the content of this information is expressly prohibited. The author, its agents and representatives shall not be responsible for any claims, actions or damages which may arise on account of the unauthorized alteration of this information.